AF560678

UNDERSTANDING BIOPRESERVATION

UNDERSTANDING BIOPRESERVATION

By

Dr. Lata Bhattacharya

Professor
School of Studies in Zoology & Biotechnology
Vikram University
Ujjain (M.P.)
(India)

DISCOVERY PUBLISHING HOUSE PVT. LTD.
NEW DELHI-110 002

First Published-2010

ISBN 978-81-8356-510-3

Published by:

DISCOVERY PUBLISHING HOUSE PVT. LTD.
4831/24, Ansari Road, Prahlad Street,
Darya Ganj, New Delhi-110002 (India)
Phone: 23279245 • Fax: 91-11-23253475
E-mail: parul.wasan@gmail.com
info@discoverypublishinggroup.com
Website: www.discoverypublishinggroup.com

Printed at:

Sachin Printers
Delhi

Preface

The present title "Understanding Biopreservation" has been written for those students interested in careers in diverse fields of biological sciences. It provides a structured approach to learning by covering all the important topics in a uniform, systematic format. The book has been comprehensively designed incorporating recent advances in this fast moving field. It also provides accessible information on biopreservation in compact form for undergraduate students in biology and related life sciences. It is intelligible to the educated layman, though it deals with some complex ideas. It is an adequate text for all the requirements of students in this area. In addition, busy lecturers who require a quick reference compendium will find it useful, particularly for tutional planning. Simple, yet hopefully clear figures and tables are provided throughout the book.

The over-riding goal of this book, and indeed of the whole *Understanding series*, is to present the essential information concering biopreservation in a compact, readily accessible form which leads itself to student learning and revision. The convergence of various approaches has generated a rich panorama of detail, the significance of which we are still attempting to unraval. The present text has been written as an introduction to this rapidly growing field.

To make the work more comprehensive and informative, the author has consulted many authoritative books, research journals, abstracts, monographs etc., so there can be no claim to originality except in the manner of treatment.

The author expresses his thanks to his friends and colleagues whose continue inspirations have initiated him to bring out this book.

The author expresses his gratitude to Mr. Wasan and staff of M/s Discovery Publishing House Pvt. Ltd. for their whole hearted co-operation in the publication of this book.

In the mean time, the author will remain sincerely responsible for any shortcomings of the book and be grateful to the readers for their suggestions and constructive criticism for the continuous betterment of the book. He takes this opportunity to appeal to the readers to send their suggestions straightaway to his Publisher.

Author

CONTENTS

1

Introduction

The target of the action of low temperatures is the water contained in large amounts in the biological objects, thus the phase transitions of water are of primary importance. A typical cryosurgery operation lasts between several minutes and an hour. Cryoaction in medicine consists of a single or several (in cryosurgery) cycles "freezing–exposition–thawing". The interval between the cycles in cryosurgery considerably increases the destruction of the living tissues, leaving the tissue for a longer time in the hypotonic state and providing the sufficient time needed for the blood microcirculation failure to develop.

Cryoaction could be modulated in time (the so-called "*dynamic*" *cryosurgery* or *cryocycling*); as experiments show, the frequency of the temperature oscillations, caused by cryoprobe, is conserved throughout the entire region, while its amplitude decays with distance.

Amorphous Ice

The possibility of a fast transfer of the object to the glassy state without crystallization (*vitrification*, from Latin vitri – glass; B.J. Luyet used also the term "amorphization") is determined by the ability to attain the high enough cooling rate to pass through the ranges of the crystallization temperatures within the time interval shorter than that is needed for the formation of ice crystal nuclei, thus vitrification is easier to accomplish for a small object.

Usually, vitrification of the biological objects is performed in the presence of the *cryoprotective agent* (CPA). Type of the cryoprotectant, its concentration, as well as the protocols of the tissue saturation and of a removal of the CPA influence the outcome of the cryoaction. Partial vitrification could occur during the conventional cryopreservation when the residual solution becomes so concentrated that could vitrify in the presence of ice.

Two phases of amorphous ice are known for a long time – the *low-density amorphous* ice (LDA), also called the *amorphous solid water* (ASW) or the *hyperquenched glassy water* (HGW), and the *high-density amorphous* ice (HDA); it is under discussion whether the very-high-density amorphous ice (VHDA) that could be obtained from the HDA by heating at the high pressure should be considered as the "true" independent phase. All three phases have a tetrahedral structure of the hydrogen bonds net, but the HDA (VHDA) has one (respectively, two) additional molecules between the first and the second shells of neighbors. Numerical experiments show the possibility of an existence of the fourth phase of amorphous ice.

Recent studies indicate that the HDA is a metastable phase while there are only two stable phases – the LDA and VHDA. A possible porous structure of the different phases of amorphous ice (and, hence, the ability to absorb and retain gases) is of great importance for physics of the extraterrestrial objects – the comets, the satellites of the outer planets, and the interstellar dust. The phase transitions from the vapour, the liquid, and the hexagonal ice I_h into the amorphous ice as well as the transitions between the different phases of the amorphous ice are considered in detail in the recent reviews by Debenetti and Loerting and Giovambattista. On the relation of the polyamorphism of water to the interpretation of the freezing phenomena in the solutions, see the paper by O. Mishima.

The term "order parameter" frequently used in description of the glassy state has a somewhat different meaning than in physics of critical phenomena and phase transitions; it is even proposed to use "order" parameter in the issues relevant to the glassy state to avoid confusion. The order parameters are related to the analysis of the rare transitions between the stable or metastable states in the free energy surface. Various activated events could be described in terms of the transitions between stable (global minimum) and metastable (local minima) basins in the free energy landscape separated by the free energy barriers (transition states) as a set of paths connecting the relevant basins. The order parameters are useful for the classification of the different metastable states accounting for their characteristics such as symmetries associated with different phases and understanding of the mechanisms of transitions analyzing the evolution of order parameters along the paths in the free energy landscape as well as of the equilibrium properties of the system dependence on the relation of the free energy to order parameters. The latter relation could be obtained in two ways:

1. By construction of phenomenological theories of phase transition (*top-down approach*); the free energy functional is usually constructed using symmetry considerations;
2. By the coarse graining (*bottom-up approach*) of the microscopic Hamiltonian.

"Physical" order parameter is indeterminate at the critical point (the point of the phase transition); transition to the glassy state is not a genuine phase transition, but a dynamic phenomenon and the glass transition temperature is not strictly defined. The meaning of the order parameters in the glassy state context is rather just a set of variables that define the properties of the glassy state. There is a discussion on how many such order parameters are sufficient for the complete description of glass, which was started in the middle of the 20th century with the appearance of the theory by R.O. Davies and G.O. Jones (that idealized the glass transition by treating it as a true phase transition) and is not finished yet. The question "how many" could be reformulated to the constraints on the so-called Prigogine-Defay ratio Π defines as

$$\Pi = \frac{\Delta c_p \Delta \kappa_T}{T_g (\Delta \alpha_p)^2},$$

where Δc_p is the difference between liquid and glass isobaric specific heat per unit volume at the glass transition temperature T_g, $\Delta\kappa_T$ is the liquid–glass difference of isothermal compressibilities, and $\Delta\alpha_p$ is the liquid–glass difference of isobaric thermal expansion coefficients.

It should be noted that the Prigogine–Defay ratio Π is not strictly defined since, first, the parameters entering this criterium are usually obtained by extrapolation to the glass transition temperature that is not strictly defined itself, and, second, the glass state relaxes continuously making its properties a function of time. One of the approaches to the description of glass transition consists in freezing-in of one or more order parameters related to the internal structural degrees of freedom involved in vitrification.

The inequality $\Pi \geq 1$ follows from the Davies–Jones theory, turning into the equality if there is exactly one order parameter; the case of the single parameter also implies that disturbances should decay to the equilibrium exponentially.

Experiments show that for the most glass-forming liquids, $2 < \Pi < 5$ and the relaxations almost always are nonexponentials, seemingly supporting the need for several order parameters. However, the recent study of this issue by N.P. Bailey et al. reported in the just cited paper (the authors analyzed the frequency-dependent thermoviscoelastic response functions of the metastable liquid) showed that properties of many liquids, in particularity those in which intermolecular interaction are dominated by the van der Waals forces, could be determined by the single order parameter (no claim is made that the complete description of the molecular structure is provided by this parameter), while the description of the properties of the hydrogen-bonding liquids such as water requires more parameters.

The process of the amorphous ice aging is of more interest for cryobiology, since it defines the conditions and longevity of the storage of the vitrified biological objects. The most important aspect of aging is the formation of critical embryo/nucleus in the metastable supercooled liquid and their extinction upon heating. Experiments on the supercooled organic glass-forming liquids showed that generation of crystal nuclei is possible at temperatures over a hundred degrees lower than the glass transition temperature.

The amorphous ice by the water vitrification was obtained a half century later than by the deposition from the vapour phase on the cold substrate at low pressure by the method resembling modern molecular beam epitaxy used to grow the semiconductor heterostructures. Crystallization in the biological objects differs greatly from the growth of the inorganic semiconductor crystals for a number of reasons: the anomalous properties of a water itself, the effect of the hydrophobic and hydrophilic biological surfaces on the water behavior, and the peculiarities of the crystallization in the heterogeneous media.

Water

First of all, pure water in both liquid and solid phases (the last known phase of water, ice XII, was discovered in 1998) possess a number of anomalous properties; a complete list of them contains over 50 entries. The anomalies manifest themselves more strongly when the water is supercooled (the term *supercooled* is used more frequently than the more exact one *undercooled*).

Water has uncommon properties also as a solvent – for example, the mobility of the solvated ions nonmonotonically depends on the radius of the ion and do depend on the sign of the charge, in contrast to continuum theory of the ion behavior in the solution.

Probably the most known and certainly the most important water anomaly is that it is liquid at the normal conditions, in contrast to many similar small molecules. Water molecule itself is "an unremarkable small molecule". All water anomalous properties are defined by interactions between water molecules. Perhaps the simplest explanation is of its liquid state at the ambient temperature: unusually high strength of interaction between two water molecules (about 20 kJ mol^{-1} is that is an order of magnitude larger than the typical thermal fluctuation at room temperature ($\kappa T_{room} \approx 2$ kJ mol^{-1}). In contrast, many other small molecules interact via the much weaker van der Waals forces (the typical energy of the interaction is of the order of κT).

Most of the anomalies of the water originate from the ordering of the water molecules in the different states (the ice crystals, the liquid water, and the gas clathrates); this order, in turn, is just a manifestation of the cooperative effect of the structurally and dynamically nonuniform net of hydrogen bonds.

The thermodynamic water properties are determined first and foremost by the static inhomogeneities - a local structure, namely, the fourfold (in the average - computations using the different interaction potentials show a considerable number of the water molecules having the coordination number from 3 to 6) net coordination - the organization of the water molecules in tetrahedra. The angle between H–O bonds in the water molecule (HOH angle) is well established to be 104.52°, which is close to both the tetrahedral angle (circa 109.5°) and the internal angle of the pentagon (108°), providing thus the ability of water molecules, in addition to tetrahedral structures, to form near-planar pentagonal ring structures.

The standard simple model of the water molecule having two regions of positive charge close to the positions of the hydrogen nuclei and two regions of negative charge ("lone pair electrons") tetrahedrally displaced with respect to the positive charges is, as high quality quantum mechanical computations show, an oversimplification. The negative (lone-pair) regions appear to be much closer to the molecular center than the positively charged regions to the hydrogen atoms and the overall charge distribution is near trigonal rather than tetragonal. The molecule's repulsive core is nearly spherical; however, deviation from the sphericity may be important in the assemblies of the water molecules. If a water molecule is surrounded by four hydrogen-bonded neighbors, in two of these interactions the central water molecules act as hydrogen-bond donor by directing its positively charged hydrogen regions to negatively charged regions (lone-pairs) of each of the two neighbor molecules. In other two interactions, the central molecule acts as a hydrogen bond acceptor of two neighbor water molecules pointing their hydrogen regions towards the lone-pairs of the central molecule. Water seems to be unique in having this donor:acceptor ratio equal to 2:2, which is considered as a very important attribute with respect to the chemical and biological consequences.

A cause of the tetrahedral near order is the sp^3 hybridization of the electron orbits. The dynamical properties - both translational and rotational - are traced down to the defects of the hydrogen bonds' net.

Both experiments and computer simulations show that in the supercooled liquids in some transient regions, relaxation times greatly differ from that in other fluid regions. In the early 1960s, G. Adam and J.H. Gibbs postulated the existence of cooperatively rearranging regions (CRR) whose molecules moves essentially independently of the rest of the system. M.G. Mazza et al. studied this issue in detail using MD simulation of the system comprising 1,728 water molecules interacting via the extended simple point charge potential (SPC/E model) for the temperature range from 350K down to 200 K. Computations have been used to extract a rotational mean square displacement

$$\langle \varphi^2(t) \rangle \equiv \frac{1}{N} \sum_l (\varphi_l(t) - \varphi_l(0))^2$$

and a rotational diffusion coefficient

$$D_R \equiv \lim_{t \to \infty} \frac{1}{4t} \langle \varphi^2(t) \rangle.$$

The computations showed that rotational dynamics is spatially heterogeneous. The study of the temporal evolution of the clusters containing more easily rotating molecules compared with that in the bulk (the authors called them *rotational heterogeneities* (RH)) and the radial distribution function of oxygen atoms within HR, which show the amplitude of the first peak that is greatly enhanced in comparison with bulk water, showed that there is a strong tendency for the more mobile molecules to be neighbors. These molecules, as a rule, have more than four neighbors and thus represent "*defects*" in the tetrahedral network of hydrogen bonds. The translational heterogeneities, while not identical to RH, certainly have similar features and the same nature.

The net structure changes significantly in the case of the spatial confinement - for example, for a quasi-one-dimensional water in the single-wall carbon nanotube (SWCN), the coordination number is less than 2. The ordering in the liquid phase is based on the structures similar to the hexagonal ice I_h – the water hexamers $(H_2O)_6$, which can form a number of configurations with close energy; the pentamers $(H_2O)_5$ observed in the clathrates are grouped into the dodecahedra. Dodecahedral water clusters have also been reported for protein surfaces. Using an elementary cell containing 14 molecules, an icosahedral structure containing 280 molecules could be constructed that can transform between the two configurations with the different density without breaking the hydrogen bonds. This model was developed by combining sheets of boat-form and chair-form water hexamers that are present in the crystallographic lattices of hexagonal and cubic ice.

Water displays a remarkable combination of rigidity and flexibility: while the hydrogen bonds network seems to be "*rigid*" with respect to the thermal fluctuations at normal temperatures, the large number of coordination defects and water's ability to vary its hydrogen bonding and participate in cooperative motions provide high molecular mobility and anomalously high proton conductivity in liquid water. The first explanation of the latter was suggested two centuries ago ("*Grotthuss mechanism*") using the hydroxonium ion H_3O^+ as a mediator agent in the proton transport; modern experimental evidence is in conflict with this simple scheme; mechanism of the proton conduction based on the ab initio computations involving the hydrogen-bond breaking process as a possible rate-controlling stage of diffusion is considered in detail by D. Marks et al. It should be noted that the high diffusivity of protons in the liquid water is relevant to some biomolecular process; also, it is the basis of the hydrogen energy – fuel cells that directly convert a fuel source energy to electricity heavily rely on this property.

In normal liquids, density and entropy fluctuations decrease with the temperature while in the water, due to the local tetrahedral geometry of the molecule interactions, these fluctuations increase. Quantitatively this behaviour could be described with the spatial (tendency of the molecules to adopt preferential separations) and orientational (deviation from the local tetrahedral arrangement) order parameters that both decrease as the volume is decreased largely due to the increased dispersion in both bond angles and bond lengths. The net of hydrogen bonds serves as a cornerstone for some water models: it is considered as a set of space filling curves; the temperature dependence of the water properties is expanded as a series in the structure functions that describe the equilibrium net of hydrogen bonds (the first function gives an average number of bonds per molecule, the second one describes the deviation from the tetrahedral structure, etc.).

The degree of the water ordering increases (respectively, decreases) when cosmotropic (chaotropic) ions are present. The character of the ion effect on the water depends on the ion radius and the charge sign and manifests itself, *inter alia*, in the viscosity η dependence on the concentration c

$$\frac{\eta(c)}{\eta_0} = 1 + A\sqrt{c} + Bc,$$

where the Jones–Doyle coefficient B is positive for the cosmotropic and negative for the chaotropic ions. On the experimental study of the effect of the polar, nonpolar, and charged molecules on the water structure see the paper by D.T. Bowron. The ordering of the water molecules increases the water thermal conductivity, making it close to the thermal conductivity of ice.

The *trans*-conformation of the hydrogen bond due to the additional interactions of protons with the nondivided electron pairs that does not participate in the forming of the bond in question proved to be stronger than the *cis*-conformation. As a rule, two OH groups of the same molecule participate in hydrogen bonds of different strength. A model of the continuous net of hydrogen bonds could account for both the strong and the weak bonds.

The presence of the far order in water is confirmed by the shape of the radial (pair) distribution function $g(\mathrm{r})$, found numerically in MD computations or experimentally by scattering of the neutrons or the X-rays. These methods are the best currently available means to probe the atomic structure of aqueous solutions. The distribution functions $g_{\alpha\beta}(r)$ of pairs of atoms α and β of the systems are derived using Fourier transformation from the neutron or X-ray scattering patterns obtained experimentally. Knowledge of these functions allows to determine atomic correlation distances, coordination numbers, and the extent of the local order around a particular atom type. An accuracy of the measurements could be enhanced using the so-called *isotopic substitution* to enrich the experimental information by exploiting the differences between the scattering patterns of isotopically labeled samples.

Both the distribution of atoms and of the interstitial voids are of interest. The *difference* distribution function $G(\mathrm{r}) = r^2 (g(\mathrm{r}) - 1)$ is well suited for visualization of the large distance correlations. The orientation-correlation function $g(\mathrm{r},\Omega_1,\Omega_2)$ could not be found directly from measurements and requires post-processing by the method of empirical potential structure refinement (EPSR) that invokes the Monte-Carlo procedures. The total correlation function $g(r, \Omega_1, \Omega_2, \Omega_{12})$ includes the polar coordinates of the vector connecting the centers of the molecules. It is proposed in the just cited paper to analyze the coordination shells using the correlation functions for the spherical layers $R_{min} < r < R_{max}$. While the radial distribution function is routinely obtained in simulations, the determination of the orientational distribution function that depends on bond angles for each distance between molecules is rather labor-intensive, thus approximations to it are of great interest.

The water structure could be described with the aid of integral equation as was pioneered in 1914 by Dutch physicists Leonard Salomon Ornstein and Frederik Zernike. Their idea was to reformulate the equation for the total correlation function $h(r_{12}) = g(r_{12}) - 1$ (g is the radial distribution function), that is, the measure for the "influence" of molecule 1 on molecule 2 at a distance r_{12} by splitting this influence into two parts, direct and indirect. The direct one is given by the so-called direct correlation function $C(r_{12})$, while indirect contribution to the "influence" is weighted by the density and averaged over all the possible positions of particle 3.

Method of Ornstein–Zernike (OZ) integral equations is especially useful in combination with the so-called central force models within which the water is considered as a simple mixture of the oxygen and hydrogen atoms and no distinction is made between the atoms interactions within the same molecule and for the atoms belonging to the different molecules. One is thus relieved from the necessity to describe the orientational properties of the hydrogen bonds and to monitor the orientation of the water molecule. The OZ equation for the pair correlation function of the spatially homogeneous system reads as

$$\gamma(r_{12}) = h(r_{12}) - C(r_{12}) = \rho\int C(r_{13})h(r_23)\mathrm{d}r_3 \qquad ...(1)$$

where

$$h(r) = g(r) - 1 = \exp\left[-\frac{\Phi(r)}{\kappa T} + \omega(r)\right] - 1. \qquad ...(2)$$

Here $g(r)$ is the usual radial correlation function defined as

$$g_2(r_1, r_2) = V^2 \iint \frac{1}{Q_N} \exp\left[-\frac{U_N(r_1, r_2, \ldots r_N)}{\kappa T}\right] \mathrm{d}r_3 \ldots \mathrm{d}r_N, \qquad ...(3)$$

where Q_N is the configuration integral and U_N is the potential energy of the system, $\omega(r) = \omega(r; T, \rho)$ is thermal potential, and $\rho = V/N$

is the density. In the case of central forces, $g_2(r_1,r_2) = g(|r_2 - r_1|) = g(r)$.

The OZ equation (1) is exact. However, its analysis requires the introduction of approximations arising from the generally unknown relation $C = C(h(r))$ that could be reformulated as finding of the so-called bridge-potentials $B(r)$. The latter are expressed as an infinite series of the irreducible diagrams. The problem of finding approximation for the bridge-potential is especially complex if long-range electrostatic interactions are present in the system. If, however, an approximation for the radial correlation function is found, all macroscopic parameters could be relatively easily determined.

A recent review by Sarkisov considers both the phenomenological models of water and the models based on the integral equations for the correlation functions of the Ornstein–Zernike type.

The computations indicate that a first-order phase transition "liquid–liquid" is possible at low temperatures. This transition is analogous to the one observed in phosphorus that has a similar structure, metal–metalloid melts (Ni–P, Fe–P, Ni–B), some metal alloys (Ni–Zr), presumably, in solutions of some macromolecules, and some other fluids. The liquid–liquid transition is also predicted to occur at high pressure in BeF_2. This transition coincides with a violation of the Stokes–Einstein relation

$$D = \frac{\kappa T}{6\bar{\omega}\eta R}, \qquad \ldots(4)$$

where D is the diffusion coefficient, η is the fluid viscosity, and R is the hydrodynamic radius of the particle. Since the determination of the viscosity from MD computations is not straightforward, the authors of the just cited paper used instead in their study the time τ_α of the so-called α-relaxation: this variable exhibits the same temperature dependence as η.

It is known that both the Stokes–Einstein relation (4) and its analog for the rotational diffusion – the Stokes–Einstein–Debye relation

$$D_r = \frac{\kappa T}{8\bar{\omega}\eta R^3}, \qquad \ldots(5)$$

that were derived by the combination of classical hydrodynamics (Stokes law for flow around the sphere) and kinetic theory – hold for low-molecular-weight fluid for the temperatures significantly higher than the glass transition temperature T_g: $T \gg T_g$. When liquid is deeply supercooled, these relations greatly overestimate the diffusion coefficients.

M.G. Mazza et al. have established that origin of this breakup could be traced down to the dynamic heterogeneities in the supercooled liquid that does not become a glass in the spatially homogeneous fashion. The authors also found that degree of the violation of the Stokes–Einstein (4) and the Stokes–Einstein–Debay relation (5) varies with temperature, since following from these relations a temperature-independent ratio

$$\frac{D}{D_r} = \frac{3}{4R^2}$$

does change with supercooling.

In the *low-density liquid* (LDL) an open locally ice-like network of hydrogen bonds is present, while in the *high-density liquid* (HDL) the local tetrahedrally coordinated hydrogen bond network is not fully developed; the structures of LDL and HDL are similar to those of the corresponding phases of amorphous ice LDA and HDA.

The behavior of the disordered media under the temperature variation is usually attributed to their ability to form stable structured objects using the directional bonds. It is turned out, however, that the directional character of the bond is not the necessary feature: the numerous water anomalies could be reproduced assuming the tetrahedral (sp^3) coordination described by the spherically symmetric interaction potential with the soft core. Such potential (ramp potential with two characteristic lengths - an impenetrable hard core and a penetrable soft core) is called sometimes the Jagla potential and the corresponding fluid, displaying many water-like structural, dynamic, and thermodynamic anomalies, called the *Jagla fluid*.

Zanotti et al. tested the hypothesis on the two phases of liquid water ("liquid–liquid critical point hypothesis") experimentally. Using the *differential scanning calorimetry* (DSC) and the neutron spectroscopy, the authors observed two states of the supercooled water differing in density and the first-order transition between these states. The existence of this critical point could explain, inter alia, the anomalous increase of the isothermal compressibility and isobaric specific heat on cooling.

Experimental study of the water molecule dynamics in SWCN showed that at 228 K the temperature dependence of the relaxation time is changed from the Vogel–Fulcher–Tammann law

$$\tau = \tau_0 \exp\left(\frac{B}{T - T_0}\right) \qquad \text{...(6)}$$

to the Arrhenius-type dependence

$$\tau = \tau_0 \exp\left(-\frac{E}{\kappa T}\right),$$

where E is the activation energy.

The authors assume that this moment corresponds to the first order transition from the low density ($\rho < 1.02$ g cm^{-3}) liquid water to the high density ($\rho > 1.14$g cm^{-3}) liquid water. Similar results on the change of the temperature dependence of the relaxation time were obtained for the water in the pores of 50 nm diameter in the ceolites. It should be noted, however, that dynamic water transition observed by E. Mamontov et al. in SWCN with the inner diameter of 14Å was not detected in the double-wall carbon nanotubes with the inner diameter of 16Å.

Since heterogeneous nucleation could be experimentally circumvented, the homogeneous nucleation theory gives an upper bound to the maximum supercooling. A homogeneous nucleation of the pure water is observed at –39 °C; more deep supercooling can be achieved underthe high pressure (up to -92 °C) and in nanopores (it should be noted that, on the other hand, the glassy state of the water in the pores and at the silica surface of Vycor could exist at the room temperature); the water dynamics corresponding to a liquid state is observed at low temperatures in the carbon nanotubes. When the temperature is lowered further, tube-like solid structures are observed named by the author's ice-nanotubes (ice-NTs) or a square-ice sheet wrapped into a cylinder inside the carbon nanotube.

The water is not unique in rather high degree of supercooling that can be achieved: the ratio of the homogeneous nucleation temperature to the freezing temperature at normal conditions is about 0.85 for the water while this ratio is 0.79 for phosphorus and 0.75 for ammonia.

Note that the value of –39 °C as the homogeneous nucleation limit below which even very small droplets would crystallize is in some contradiction with experiments by L.S. Bartell and J. Huang, who observed freezing of droplets formed by condensation of water vapour in the supersonic Laval nozzle to cubic ice I_c at temperatures close to 200 K.

Recently the possibility of the existence of ice at room temperature has been established. Computations performed by A. Wissner-Groos and E. Kaxiras showed that a layer of diamond coated with sodium atoms will keep a thin layer of water in the frozen state up to 108 degrees of Fahrenheit (42 °C). This discovery (if confirmed by experiments) will greatly advance biocensors and other implanted devices, since an ice layer covering the nonorganic surface will smooth the diamond surface and prevent clotting proteins from attaching the surface.

The electric field of strength about 10^6 Vm^{-1} raises the crystallization temperature of the water in the nanoscale slit to room temperature. Both the restrictions on the water molecule's mobility in the normal to the wall direction and ordering of the dipoles in the electric field help the formation of the stable net of hydrogen bonds and of the critical size embryo. This effect is reversible - decreasing the field strength results in the ice melting.

S. Wei et al. studied experimentally the effects of dipole polarization of water molecules on ice formation under an electrostatic field with strength E in the range 10^3–10^5 Vm^{-1}. When a water molecule is in the external electrostatic field, there is force moment M acting on the molecule given as $M = \mu_0 E \sin\theta$, where μ_0 is the electric dipole moment of the water molecule and θ is the angle between the dipole moment direction and the field direction. The water molecules, excluding the case when the dipole moment is parallel to the field direction, tend to turn in the direction of field to reach the stable state. The potential energy of the interaction U is $\mu_0 E \cos\theta$. Assuming the Boltzmann distribution law, the distribution of water molecules could be written as

$$f(U) = A\exp\left(-\frac{U}{\kappa T}\right) = A\exp\left(-\frac{\mu_0 E\cos\theta}{\kappa T}\right).$$

Evidently, the Boltzmann distribution function attains maximum when all water molecules have their dipoles oriented along the electrostatic field direction that assists in forming the critical nuclei. The authors, however, found that in contrast to the nucleation, the growth time was unaffected by the presense of the field. The effect of the electric field on the nucleation of the ice crystals is known for over one and a half centuries. The electric field could be used to assist the formation of the ice nuclei in the water solution at the given temperature to control the size of grains in the growing polycrystal or the liophyllization time.

Liquids can exist in metastable supercooled state; in contrast, solids cannot be superheated since in contrast to freezing there is no energy barrier for melting. The temperature of solids is always below their (pressure-dependent) bulk melting temperature T_m. When solid temperature is close (from below) to T_m, the phenomenon of premelting occurs: mobile liquid is observed on at one facet of a wide class of materials that includes water, solid rare gases, semiconductors, metals.

The premelting is subdivided into the surface melting (at the solid–vapour interface) and the interfacial melting (at the interface between the solid and the foreign substrate). The existence of the interfacial premelting depends on a competition between attraction of the liquid to the solid (called *adhesion*) and attraction of the liquid to itself (*cohesion*). The film thickness d is proportional to the deviation of the temperature from its bulk melting value:

$$d = \lambda (T_m - T)^{-\frac{1}{\nu}},$$

where the parameter ν characterizes the fall-off rate of the intermolecular forces and equals $\nu = 2$ for the electrostatic and $\nu = 3$ for the van der Waals forces. It could be seen that the film thickness increases rapidly as the temperature tends to the bulk melting temperature, that is, the crystal melts from its surface inwards.

Ice exhibits all aspects of premelting, including grain-boundary melting at junctions between the individual crystals in the polycrystal. Polycrystalline ice contains liquid water due to the impurity and curvature depression of the freezing point in microscopic channels (a few tens of micrometer wide) called veins formed by the tri-grain junctions. Water is also present at nodes separating four grains. Water forms a network distributed thoughout the whole polycrystal.

The so-called "*growth-melt asymmetry*" was observed for pure ice crystals at high pressure (2,000 bar). As the pressure increases, the melting temperature decreases and the system passes through a roughening transition with change of the cylindrical shape into the hexagonal prism due to the formation of the facets. Both growth and melt shapes are hexagonal, but the corners appear to be rotated by 30°. This behavior is explained by the more fast melting of the initial corners where the molecules are weakly bound; the rate of the corners' recession is considerably larger than that of the facets.

Some of the water anomalies are also exhibited by other tetrahedrally coordinated or simply tetrahedral liquids with different bonding varying from the purely ionic nature in the case of BeF_2 to the purely covalent in the case of SiO_2.

Biological Water

Second, the behavior of the water molecules near the hydrophobic or hydrofillic surfaces cardinally differs from their behavior in the bulk. The terms *bond*, *ordered*, *biological* water are used, and the cryobiologists also speak about *unfreezable* water (it is more correct, however, to use the word *unfrozen*, allowing for a nonequilibrium state). It is possible to distinguish *buried* or *inner* water that is held by the hydrogen bonds in the inner hydrophobic regions of the biological molecules and sometimes form clusters. Such water is present in most globular proteins; in spite of the permanent exchange with the exterior molecules with the characteristic times from 10 ns to 1 ms, it should be considered as an intrinsic element of the protein. The term *vicinal* water (from the word vicinity) is usually used in respect to inorganic surfaces.

The importance of the water for the stability and function of biological molecules hardly can be overestimated – the water itself could be considered as the biomolecule and even is called the 21st amino acid. In some cases the normal behavior of the protein molecule is possible only if the water on its surface forms (in terms of the percolation theory) an infinite cluster. The hydrogen bonds from the water to the peptide groups are only slightly stronger than the hydrogen bond between the water molecules. MD computations reproduce the anomalous behavior of the water near the surface.

To establish the hydrogen bond with the protein surface water either donates its proton to protein (to carbonyl/carboxyl and hydroxyl oxygen atoms) or accepts proton from the protein (from amine, amide, and hydroxyl groups that are present in the protein backbone (NH) and in some side chains (NH, NH_2, NH_3)). Water molecule also could mediate interaction between two protein atoms ("*water bridge*"). Hydrogen bond contribution to the solvation energy was determined recently by M. Petukhov et al. for several atom types (NH/NH_2-uncharged groups, H in hydroxyl groups, NH_2/NH_3-charged groups, O-uncharged carbonyl/hydroxyl groups, and O-charged carboxyl groups).

The change of the average orientation of the water molecules near the interface could be assessed by the normalized probability distribution for the angle between the water dipole and the normal to the interface and also for the angle between the normal and the vector joining the hydrogen atoms. In addition to the water/solid interfaces of different nature, the water/carbon tetrachloride interface was extensively studied, since this system serves as a prototype for hydrophobic interactions and in many respects mimics the biological membrane.

The variation of the water properties near the biological surface is caused by the regrouping of hydrogen bonds – by ordering of the water molecules – determined by both the nature of the surface (polar or non-polar) and the spatial restrictions on the structure of the hydrogen bonds net (cutoff of the linear correlation scale); the mobility of the water molecules and the water dielectric constant considerably decrease, the water viscosity increases. The effect of the wall proximity on the crystallization temperature is known from observations on the soil freezing and the water behavior in the pores of minerals. Zangi simulated using MD method the water crystallization in the narrow (several diameters of the water molecule) gap for the case of the inert surface that does not form the hydrogen bond with the water molecule. P. Kumar et al. studied the effect of the interaction potential for the gap containing from two to three layers of the water molecules. Two different interaction potentials were considered: Lennard-Jones and purely repulsive potential. The increase of the pressure shifts the water molecules towards the hydrophobic surfaces and increases the value of the tetrahedral parameter; no pressure effect is observed for the case of hydrophilic surface.

MD computations using the density functional theory show the existence of the thin water layer of the higher density near the hydrophilic surface and the significant increase of the strength of bonds between the water molecules in the first hydration shell near the nonpolar surface in comparison with the molecules in the neighbor shells. The authors studied the ordering of the water molecules in the hydration shells of benzene and cyclohexane. They stress that correct classification of surrounding water molecules into discrete shells could be done either by using a distance cut-off, measuring the distance between the water oxygen to the nearest solute carbon atom, or by Voronoi tessellation of the simulation domain, the latter being more robust. Atoms in contact could be explicitly defined as those sharing a face of their surrounding polyhedra accounting for that fact that being closer than some specified distance is not sufficient to be in contact.

Zewall investigated the evolution of the water structure on the model amphiphillic surface (the specially processed silicon surface was used) exploiting the superfast electron crystallography and found an order of magnitude difference between the characteristic times of the decay of the hydrogen bonds structure for the surface layer and in the bulk. On the other hand, the measurements of the rotational mobility of the water molecules in the protein hydration shell and in the bulk using the method of the magnetic relaxation dispersion give only twofold reduction.

Anomalous water properties also manifest themself in the behavior of the fluid viscosity in the layer confined between two close planar surfaces. If the gap thickness is greater than 8–10 molecular diameters, no effect is observed either for the water or for the standard organic liquids. For the gap between five and eight molecular diameters, however, the effective viscosity increases greatly in the case of normal liquids with much smaller relative change for the water.

Computer simulations show that the changes in the dynamic behavior of the water near the surface depend on the morphology of the latter: while the water dynamics typically slow down near the nonattractive *rough* surface, the dynamics speed up near the nonattractive smooth surface.

The direct proofs of the anomalous dynamic properties of the *biological* water were obtained in the experimental study of the water near the lactose molecule using the absorption spectroscopy in the terahertz range (with the time resolution about a fraction of picosecond) and the water at the protein surface with the femtosecond time resolution; several layers of the water ordered due to the interaction with the heads of lipid molecules are also observed near the surface of the biological membrane.

The water molecules adjacent to the lipid bilayer interact strongly with the headgroups of the lipid molecules and this interaction can extend to several molecular diameters from the membrane surface: for the case of 1,2-dipalmitol-*sn*-glycero-3-phosphoholin (DPPC) phospholipids computations based on the local density as a function of the distance from the surface show the existence of four distinct regions of water, the first two corresponding to the first and second solvation shells surrounding the phosphocholine groups. It is anticipated that these water layers can effectively influence the membrane permeability, posing an obstacle to the transport of permeant molecules to the membrane and even be a rate-limiting stage in permeation. The study of the water dynamics in the micelles showed that there is no considerable variation of the water mobility in the core while a significant reduction is registered in the Stern layer that contains the polar groups of the amphiphilic molecules and the counter-ions.

Both direct measurements by the terahertz absorption spectroscopy and MD computations of water near the surface of solute (lactose molecule) show over a hundred water molecules beyond the first solvation shell are affected by the molecule, that is, belong to the hydration layer that extends to approximately 5–6Å from the surface (123 such molecules were found in the experiment, 112 were from computations). Molecules in this hydration layer show slower dynamics in comparison with molecules in bulk. The authors assume that this retardation of water dynamics by solute is the source of the stabilization effect on the proteins and membranes in a dehydrated or a frozen state by solvated mono- and disaccharides.

On the basis of the experimental data and MD computations it was suggested to distinguish three kinds of water molecules – free and bound viz.:

1. Molecules bonded to other water molecules only
2. Molecules having a single bond with the macromolecule's polar group
3. Molecules having two bonds with the macromolecule's polar group

Somewhat more formal classification of hydration water molecules was suggested by M. Nakasako in relation to the water present in protein crystals. The four classes are "inside" (water molecules that occupy cavities in the protein molecules), "contact" (molecules that are located outside the SAS of protein and mediate interactions between adjoining molecules in the crystal; these contact-class hydration sites are formed during the crystallization process), "first-layer" and "second-layer." The electron density in the first-layer hydration shell is greater than that in bulk solvent, thus dielectric properties of hydration water molecules of this class must differ from those in bulk solvent. Hydration water molecules of the first-layer and second-layer classes form aggregates using hydrogen bonds that link together and are indirectly connected to the polar protein atoms forming a network of hydrogen bonds on the protein surface.

The reduction of the water mobility near the solid surface is qualitatively similar to the behavior of the supercooled water; sometimes even an "*equivalent temperature shift*" is determined: for example, the value of this parameter for the water in the nanogap between two smooth *hydrophobic* surfaces is estimated from MD computations as 40K, while M.C. Bellissent-Funel reported the estimated temperature shift of 20°C that is attributed to the difference life times of the hydrogen bonds. On the other hand, simulation for the case of the *hydrophilic* surface of hydroxide $Mg(OH)_2$ shows that the confinement effect is more akin to the pressure increase. When the temperature decreases, similar to the aforementioned experiments with SWCT, the change of the relaxation time dependence on the temperature is observed approximately in the same temperature range (222 K and 220 K for the water on the surface of DNA and lysozime, respectively).

The discrepancies in the dynamical data of the biological water obtained in the different measurements are related to the difficulties of separating the water and the macromolecule motion in some experimental approaches; for example, when quasi-elastic and inelestic neutron scattering is used, the elastic contribution from the biological molecule itself could be significant. B. Halle noted that reduction of the mobility of the macromolecules in the solution in comparison with the mobility of the equivalent (hydrodynamically) body (a sphere, an ellipsoid) is at least partly explained by the significant surface irregularity.

Crystallization in Heterogeneous Media

Finally, the biological media has a complex microstructure. In the absence of a seed surface, ice formation in the supercooled solution is initiated by the homogeneous or the heterogeneous nucleation. The heterogeneous nucleation dominates if the following nonequality ("*wetting criterium*") is valid:

$$\sigma^{vs} - \sigma^{vl} < \sigma^{ls},$$

where σ^{ij} is the surface tension on the interface between phases i and j, subscripts v, l, and s denote vapour, liquid, and solid, respectively. The nucleation is observed at the cell membrane or at the objects contained in cytoplasm; the importance of the nucleation on the subcellular objects increases as the temperature decreases; heterogeneous nucleation has not been observed in the cells of some hardwood plants and co-existence of ice and supercooled water is possible.

The phase transition in the biological media occurs in the finite temperature interval (in this respect biological tissues are not unique - the same behavior is observed, for example, in freezing of the soils). There are two reasons for the finite temperature interval of the crystallization in the biological media: continuous variation of the solute concentrations in the different tissue compartments of finite volume and the presence of the extracellular matrix.

The freezing process of the pure water and aqueous solutions, being cooled externally, is well known. First, the temperature decreases with time until the maximal supercooling ΔT_m is achieved. Once nucleation occurs, the rapid crystal growth results in the releasing of the latent heat of fusion faster than it can be removed from the system and rise of the temperature to the equilibrium melting temperature. The system temperature then remains constant for some time until all water is transformed into the ice (the so-called "*latent heat plateau*"), then (if external cooling is continued) again decreases.

The behavior of the freezing aqueous solution displays a few differences:

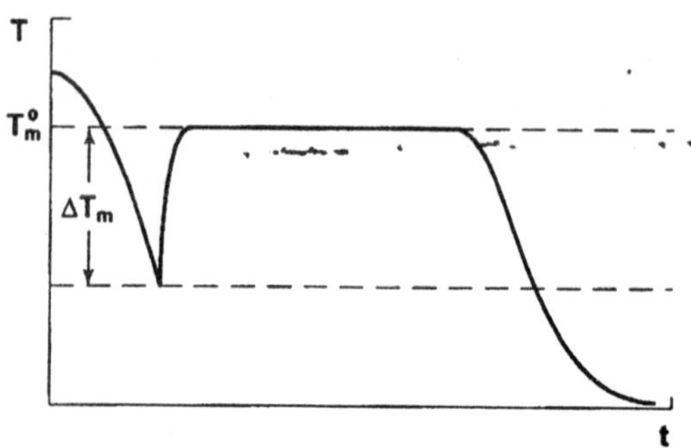

Fig. 1.1. Freezing of pure water.

1. The absolute temperature of supercooling, i.e. the nucleation temperature, is usually higher than that for the pure water, since the presence of solutes promote nucleation
2. The temperature rise after nucleation is smaller, since the equilibrium melting temperature of the solution is depressed
3. The temperature on the "latent heat plateau" is not constant, but slowly decreases due to the continuous growth of the solute concentration and, hence, continuous lowering of the equilibrium melting point of the solution
4. There is another peak on the temperature vs. time curve related to the heat release at the point of the eutectic crystallization; cooling below T_{Eu} will be continued only after the solution has solidified completely.

The behavior of water in porous media, including freezing, is known to differ from the behavior of bulk water. The relative strength of the fluid–wall interaction to the fluid–fluid interaction determines the freezing point variation: the shift $T_{f,pore} - T_{f,bulk}$ is positive if the

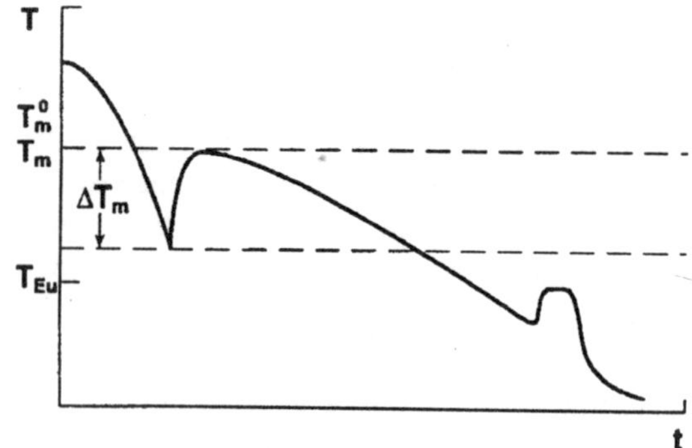

Fig. 1.2. Freezing of solution.

fluid–wall interactions are more attractive than the fluid–fluid interaction and negative otherwise.

Sometimes it is assumed that mechanical damage due to freezing in the porous material is attributed to the volume expansion at the phase transition in capillars. In fact, the mechanism could be more involved since freezing in the capillars themselves is suppressed by a significant depression of the freezing point when the diameter is small and the curvature of the crystallization front is large. Ice crystals are first formed in the relatively large voids and are called ice lenses. These crystals are not insulated and the water is supplied through the capillars to support further ice growth. Finally, large enough crystals act mechanically on the porous matrix causing its deformation. In the case of wet soil freezing, this phenomenon is known as frost heave. It alters landscapes, destroys roads, runways, pipelines, building foundations in countries with severe winters, and redistribute methane in the permafrost regions.

2

Basis of Preservation

There is a growing interest in and demand for improved methods of cell and cellular-construct preservation with the requirement for rapid return to prepreservation, normophysiological function. Implicit in this need is the development of methodologies that protect the genome and proteome thereby avoiding stress-related genetic selectivity. Unfortunately, a growing body of evidence suggests that the multiplicity of stresses attendant to many preservation protocols fails to meet these requirements. In addition, the use of animal-derived products, as common components of cell storage processes, raises concern over system contamination with nonnative protein-derived materials (i.e., prions, etc.).

Interdisciplinary efforts to advance the effectiveness of long-tried "*successful methods*" of cell, tissue, and organ preservation have now led to the development of a new scientific specialty — *biopreservation*. From a historic perspective, efforts to sustain living biologics in a dormant state supportive of "on-demand" reanimation have centered primarily on either hypothermic (refrigerated) storage or preservation under the comparatively extreme conditions associated with freezing. *Hypothermic storage* typically includes maintenance at temperatures above 0°C but below a normothermic range (32–37°C) of mammalian species. The principles that have governed advances within the hypothermic storage research focus have depended primarily on improvements in tissue and organ preservation in support of transplantation; namely, factors related to ion balance, buffering capacity, free radical scavenging, oncotic support, and the provision of nutrients. *Cryopreservation* is defined as the maintenance of biologics at temperatures below -80°C and typically below -140°C (below the reported range of the nominal glass transition temperatures of pure water). Methodological developments defined under the cryopreservation mantra are linked to the principles that emerged from Mazur's "Two-Factor Hypothesis," which relates survival of cells in solution to a dependence on the rate of cooling of the vessel (i.e., vial, straw, etc.) containing the preserved biologic. Suboptimal cooling rates in the presence of molar levels of cryoprotective agents such as dimethyl sulfoxide and glycerol cause cells to experience prolonged periods of exposure to toxic levels of these agents. The resulting "*solution effects*" are *a priori* evidence of lethal conditions, a supposition gleaned from Lovelock's classic demonstration of the relationship between survival and sodium chloride content in human RBCs. Cooling at supra-optimal rates would not afford cells the opportunity to dehydrate (reduce the levels of freezable water) such that intracellular water compartments would maintain a chemical equilibrium with the extracellular ice leading to the probability of lethal intracellular ice formation.

What is striking about the parallel developments within these research theaters is the relative isolation of the cryopreservation thought processes from the cellular fundamentals that grew from organ-based hypothermic storage research. That is, cryopreservation has focused nearly exclusively on the physical parameters associated with the water-to-ice phase change (i.e., water flux) during the preservation process at the expense of understanding that a *chill-freeze continuum* exists and that the resulting *hypothermic continuum* will impact survival. This failure to factor in the impact of the hypothermic continuum as a perturbing (stressful) condition during the application of a cryopreservation protocol has contributed to the limitations faced in today's efforts to recover normally functioning cells upon removal from cryogenic storage. It is essential, if not critical, to recognize that despite the presence of extracellular ice, cells that are *structurally preserved* (avoid intracellular ice formation) remain in a state of deepening hypothermia until reaching the vitrification state (Tg) of the preservation medium. Importantly, during this thermal excursion, solute levels continue to elevate due to freeze concentration. Cell function, while suppressed, does not cease until the intracellular glass transition (Tg) temperature is reached.

This segregation of low-temperature-directed research, when coupled with the compartmentalization of anhydrobiosis or the dry-state preservation sciences, has created a series *subdisciplinary cul-de-sacs* from which new advances have been slow to emerge. Accordingly, *biopreservation* as an integrative specialty can be defined to include those sets of processes that suppress biological aging in a manner supportive of post-preservation restoration of function and maintenance of structure. Biopreservation incorporates relevant engineering principles with developments in cellular and molecular biology including cell signaling, genomics, proteomics, metabolomics, systems biology, and computer sciences, with a strong underpinning in the fundamental principles derived from studies in anhydrobiosis and cryobiology. Importantly, biopreservation represents the simultaneous application and management of numerous — often poorly defined and not fully recognized — lethal conditions with the expectation of normal recovery. In practice, the success of this field is critical to cell-based research, related therapies, reproductive management (animal and human), banking of biologics (clinical and environmental), drug discovery, organ preservation, pest control, and other diverse applications.

Hypothermic Continuum

For mammalian systems, nearly all biopreservation stratagems begin with a reduction in temperature from 37°C to the 20–25°C range, but most commonly to 0–10°C. A maintenance target of 4°C is common and less related to any known biological rationale than to the fact that liquid water reaches its maximum density at this temperature. Cooling represents a change in the energy state of a biologic due to heat flow to the environment. In effect, the kinetic energy necessary to support the chemical reactions that define the metabolome is reduced, resulting in the uncoupling and recoupling (shunting) of biochemical reactions. These biochemical imbalances cause the depletion of adenylates due to the failure of aerobic production of ATP, which in turn disrupts membrane-mediated transport. With the progressive drop in temperature, cells experience rapid gains in calcium, the loss of intracellular potassium and gain of sodium, and intracellular acidosis with pH approaching four.

In addition to metabolic imbalances, there are measurable changes in cell and organelle membrane characteristics reported as phase changes in the lipid domains and indicated by discontinuities in the Arrhenius plots. These membrane transitions represent a structural change in the membrane from the liquid-crystalline state to the solid gel state. As a result, membranes become "leaky," thereby contributing to trans-membrane ionic imbalances. Concomitantly, there is an increase in the activation energy of membrane-bound enzymes, which compounds the problems associated with ionic imbalance, activation, and leakage of lysosomal and lipoprotein hydrolases, activation of calcium-dependent phosolipases, and the release of free fatty acids. There are also mitochondrial-related events including reduction in the bcl2-bax (anti- and pro-apoptotic proteins) ratio leading to cytochrome c release and activation of the apoptotic cascade, which if not inhibited will lead to cell death. Simultaneous with these disruptive events is further physical compromise of the intracellular milieux characterized by the disruption of the cytoskeletal matrix.

These potentially detrimental events occur with "minor" changes in the kinetic energy levels (a decrease of 10°C equals a 3% decrease in available thermal energy). One simple measure that represents an average of the "total" change in metabolism is Q_{10}, which in mammalian systems calculates to an approximate 50% decrease in oxygen consumption (metabolism) for each 10°C reduction in temperature. Accordingly, the oxygen consumption of a cell at 5°C is approximately 6% when compared to 37°C. Q_{10} represents a simplification as it does not reflect individual reactions but a differential or average of regulatory and nonregulatory enzymatic processes and hence the net of uncoupling/recoupling and shifts in metabolic pathways. Q_{10} has been observed to increase dramatically with the onset of freezing.

Accordingly, hypothermia impacts innumerable components of the cellular machinery including energy status, macromolecular reactivity and stability, adenylate levels and availability, and biological reducing power. When considered within the full context of the oxidative stressors presented to a mammalian cell in the cold, along with the generation of free radicals, hypothermia provides multiple routes to the initiation of apoptosis or gene-regulated cell death. While select mammalian species have evolved a mechanism supportive of chronic whole-body hypothermia (i.e., hibernators) or seasonally adaptive regional heterothermy, few human cells *in vivo* exhibit a genetic propensity supportive of low-temperature survival beyond a few hours.

Cooling does provide short-term *in vitro* survival primarily through the decrease in metabolism and the reduction in oxygen and nutrient demand, which work to conserve chemical energy. These processes only conserve to a point, though. Hypothermia represents far more than the apparent effects associated with the "*single variable*" of reduced temperature. Little is known about the survival-related stress responses of mammalian cells other than reports of the production of heat shock (stress) proteins (HSP), which putatively act to chaperone proteins to prevent cold-induced denaturation. Hypothermia without manipulative intervention sets the stage for progressive cell injury during each of its three phases (cooling, maintenance in the cold, and rewarming).

CRYOPRESERVATION

Cryopreservation protocols begin with hypothermic exposures, extend through the *hypothermic continuum*, and reach equilibrium in the glassy, or vitrified, state. These steps are "reversed" during the thawing process. The addition of cryoprotective compounds represents the next step in the preservation process. These cryoprotective agents include a diversity of penetrating (membrane permeable) and nonpenetrating agents typically contained within an isotonic cell culture or buffered electrolyte media. The duration of the incubation period is determined empirically but commonly lasts from 10 to 30 minutes. As temperature is lowered below the equilibrium freezing point of the preservation medium following the addition of penetrating cryoprotective compounds, nominal cooling rates are applied over user-varied subfreezing temperatures (–40 to –80°C). Following this is the transfer to ultralow temperature storage, i.e., liquid nitrogen immersion, liquid nitrogen vapor phase, or ←135°C mechanical storage. This temperature range is ideal; partly due to the ready availability of liquid nitrogen, a nonflamable cryogen, and the fact that this temperature range falls below the reported glass transition temperatures (Tg) of pure water. The glass transition temperatures for cryoprotective mixtures are reported to be in the –115 to –90°C range.

Below the glass transition temperature, the viscosity of the medium increases exponentially so that all measurable molecular translational motion is lost. While molecules retain rotational and vibrational movement, the moment of diffusion (distance equivalent to one molecular diameter) of a given molecular species increases from fractions of a millisecond to decades.

Hence the presumption that molecular interactions (i.e., metabolism) are impossible during the preservation storage interval. It should be noted that the *hypothermic continuum* ends with the glass transition state. Prior to reaching the Tg, chemical reactivity is maintained, albeit at reduced rates, yielding the potential for sustained free-radical damage. It is for this reason that long-term storage at –80°C is ill advised even for a cellular biologics such as serum, macromolecules, etc.

Following the addition of and incubation in the cryoprotective cocktail, cooling is extended at a nominal ("optimal") cooling rate (1°C min^{-1} is common with many mammalian cells). A "seeding" step (ice nucleation) is included in the -2 to -6°C range to prevent excessive under- cooling ("*supercooling*") of the cell and its bathing medium. It is advantageous to initiate the extracellular freezing process close to the equilibrium freezing point of the cryoprotective medium. Seeding supports the comparatively gradual growth of extracellular ice, which allows for the osmotic efflux of water from the cell resulting in cell shrinkage and the equilibrium freeze concentration of solutes across the cell membrane. Freeze concentration reduces the availability of freezable water in the cell while driving up the intracellular viscosity toward Tg. If cooling rates are too rapid, adequate cellular dehydration will not occur and the probability of lethal intracellular ice formation will increase.

The consequences of supraoptimal cooling are recognized upon thawing as cell rupture and early stage (0 to 6 hours post-thaw) cellular necrosis. If cooling rates are too slow, it is believed that the extensive exposure to the freeze concentrated solutes (now multimolar levels) will result in toxic "*solution effects.*" One facet of the manifestation of "*solution effect*" toxicity is the initiation of post-preservation necrosis peaking from 6 to 12 post-thaw hours, followed by a second peak in post-thaw apoptosis from 12 to 36 post-thaw hours (cell-type dependent) followed by a secondary peak in necrosis related to the interplay between apoptosis and secondary necrosis. This complex post-thaw survival profile, defined as Delayed Onset Cell Death (DOCD), poses significant challenges to post-preservation utilization of thawed cells.

Two of the more common methods of controlled rate cooling (CRC) include both microprocessor-controlled injection of the gaseous phase of boiling liquid nitrogen (active controlled rate cooling), and passive methods that employ insulated alcohol baths placed in -80°C mechanical freezers. Active CRC devices monitor a representative sample vial, straw, or bag, and adjust the inflow of cryogen to maintain a preset cooling profile. Profiles are set to maintain a standard rate of cooling over a prescribed temperature range, to "seed" (nucleate) the sample with a thermal shock administered by a surge in cryogen, and to "flatten" the temperature rebound that occurs upon the initiation of ice growth due to released latent heat of fusion of water. Active CRC also provides records of the cooling profiles. Passive CRC devices contain the sample in a central cavity surrounded by — but isolated from — an alcohol layer. When placed in a -80°C freezer, a curvilinear rate of approximately 1°C min^{-1} is obtained. Seeding is accomplished by mechanical agitation at a prescribed time (user determined) during the cooling period. No provision is made for managing the temperature rise in the sample due to the release of the latent heat.

Both approaches to controlled rate cooling provide satisfactory results despite the variability in cooling profiles. One is left to conclude that the resulting rate variations are tolerable depending on cell type (within certain constraints) and that attention to the elevation in temperature that occurs during the ice crystal growth due to the release of the latent heat is not a primary consideration. Others have demonstrated that optimal cooling rates are dependent on the initial levels of cryoprotective agent. Upon completion of the CRC step in the cryopreservation process, the sample is placed into long-term storage. Independent of the method of cooling, the cell is maintained in a vitreous state, provided that temperature does not fluctuate above the Tg of the solution. The cell will also be surrounded by a thin layer of vitrified cryoprotectant encased in the mass of extracellular ice.

Cells may be cryopreserved in an "ice-free" state by incorporating high molar concentrations of cryoprotectant mixtures added stepwise during the cooling interval. This approach to vitrification eliminates most ice from the preservation environments and may offer benefit when attempting to cryopreserve complex tissues and organs. High solute-dependent vitrification procedures do, however, expose the cells to equivalent "*solution effects*," thereby exacerbating stress factors associated with the *hypothermic continuum*.

Rapid thawing of a cryopreserved sample is preferred. It is attained by placing the sample in a 37–40°C bath, agitating until most of the ice melts, and then eluting the cryoprotectant cocktail with cell culture media in a single step. A stepwise dilution process alternatively may be used when high levels of cryoprotectant (above 10%) were used during the initial cooling. Stepwise elution minimizes the volume excursions of the cell, thereby preventing mechanical damage and rupture of the cell membrane. Mazur has provided elegant descriptions of the breadth of physicochemical changes that occur during cryopreservation.

Despite intensive research focused on improving cell preservation spanning more than a half century, the field continues to face many challenges. The following are excerpts from *Life in the Frozen State*, a text from 2004 that details the contemporary state of the cryopreservation sciences. Lane, in a chapter projecting the future of cryopreservation, states "few scientific problems have proved as intractable as cryopreservation ... cryobiology has been straitjacketed by its need to conform to the intractable laws of biophysics. For all its successes, cryobiology has been stuck in a rut." Mazur, in his introductory chapter in the same text, concludes, "the problem today is that applying basic principles of biophysics simply cannot solve many of the remaining challenges in cryobiology." Others have commented similarly. Fahy suggests that "the detrimental effects of cryoprotectants are almost as relevant to cryobiology as are their cryoprotective effects," while Fuller laments that "a unifying concept for cryoprotective agent chemical toxicity in cells remains beyond our grasp." Parks, on the other hand, writes, "the interaction observed between cooling and cryoprotective agents on cytological damage ... supports the notion that detrimental changes ... are compounded by other requisite steps in a cryopreservation protocol." Gao et al., through a specific example, emphasize this concern when they suggest, "the empirical methods of cryopreservation developed in the 1950s are still used today. The motility of post-thaw cryopreserved sperm (bovine) is usually 50% or less that of prefreeze motility and exhibits wide variability among individuals." They go on to write, however, that "the use of similar cryopreservation procedures... are much less effective..." in other species, and "the success of using cryopreserved sperm is still so low that it is not economically feasible to... use this approach."

While argumentative within traditional cryobiology, the message is gaining clarity. Studies focused on the biophysical basis of cryopreservation have yielded successful methods that prevent the formation of destructive intracellular ice. However, these emergent processes induce detrimental post-thaw effects and fail to account for the biology attendant to the cellular stress responses.

With the first reports of glycerol serving as a protective solute in a naturally freeze-tolerant organism and the simultaneous application of this discovery to avian spermatozoa and human erythrocytes (RBC), mammalian cryopreservation research began a decade of advancements that include the addition of DMSO to the preservation cocktail mix. By focusing preservation attention on two highly differentiated cellular "products" (RBC and spermatozoa) with fixed post-differentiation life-spans, the full spectrum of the impacts of preservation stress on the complex biology of normal functioning cells was obscured. In effect, these models have provided an effective "cloak" that has obscured the full spectrum of events associated with post-thaw, cryopreservation-induced, delayed-onset cell death.

As methodological developments proceeded with nonterminally differentiated mammalian cells, many cell types proved refractory to cryopreservation. Even those cells that are "successfully preserved" often demonstrate significant post-thaw death (30–70%) within 24 to 48 hours. Structural preservation is afforded to these cells, but a clear inability to manage the preservation-induced stresses is apparent.

Table 2.1 lists select stress factors encountered by a cell during the hypothermic continuum along with the known inducers of apoptosis. In addition, any of the stress factors that result in even partial physical damage can also initiate the launch of the necrotic cell death cascade or "cross talk" between the apoptotic and necrotic cascades followed by secondary necrosis. Energy deprivation, whether driven by metabolic uncoupling or by hypoxia, can initiate apoptosis. If energy deprivation is severe, however, a transition from initial apoptosis to secondary necrosis will result.

Table 2.1. Putative stress factors characteristic of the *Hypothermic continuum*

Hypothermia	*Known Initiators of Apoptosis*
Metabolic Uncoupling/Shunting	Biochemical Alterations/Inhibitions
Energy Deprivation	Energy Deprivation
Ionic Imbalances	Ionic Imbalances
Cellular Acidosis	Cellular Acidosis
Protease Activation	Protease (Caspases) Activation
Membrane Phase Transitions	Membrane Alterations
Free Radical Production	Free Radical Accumulation
Cytoskeletal Disassembly	
Freezing	
Water Solidification (Solute Concentration)	Ionic Imbalances
Cell Volume Excursions	Membrane Alterations
Hyperosomolality	Ionic Imbalances
Protein Denaturation	Biochemical Alterations

This listing of factors creates a relatively clear picture of the critical involvement played by the cell's "biology" in responding to a cryopreservation regime. Accordingly, a second-generation preservation paradigm must be developed to improve preservation outcomes.

Beyond the Red Cell Paradigm: Hypotheses Guiding Preservation Solution Development

The generic listing of stress factors may serve as a template to guide the design of improved preservation methods, assuming adequate "*structural preservation.*" There are well-noted differences in the sensitivity of various cell-types to preservation processes. Van Buskirk et al. reported on these variations in three cell types, indicating a possible need for "cell-matched" preservation solutions. The basis for there being differing kinds of cellular survival was linked to the manner in which caspase-3, an apoptotic executioner enzyme, was inhibited. This study provided the basis for the *multisolution hypothesis*, which suggests that distinct cell types manage their cell death pathways through differing molecular pathways. Hence, the requirement for unique management strategies may necessitate individualized preservation media.

The new challenge facing biopreservation is the need to adopt a molecular-based logic to develop the optimal preservation solution(s) necessary to support a cell's round-trip excursion through the *hypothermic continuum*. The earliest attempts at intelligent design of a preservation medium were made by the Belzer and Southard team when they developed ViaSpan (the University of Wisconsin solution) to support the transport of organs (pancreas, kidney, and liver) in anticipation of transplantation. ViaSpan, formulated exclusively for short-term hypothermic storage, was the first solution designed to manage select putative stress factors and became the first intracellular-like preservation medium. In the decade that followed, additional organ preservation solutions were commercially developed.

Extracellular vs. Intracellular-like Preservation Media

More recently, the issue of preservation solution design has moved beyond the simple addition of a penetrating cryoprotective agent (5–15%) to cell culture media, buffered saline, or mixtures including serum or a protein component. While these "historic designs" provide apparent structural preservation, they are often severely limiting in supporting cellular function. These limitations are apparent when assessing post-thaw survival and specific cell functions. Little cell death is noted within one hour post-thaw. However, over the next 24 hours in culture, nearly 70% of the cells succumb to cryopreservation-induced delayed- onset cell death that is unrelated to physical damage. With this "*optimized protocol,*" greater than 90% of the cells avoided ice-related damage. Cryopreservation in extracellular-like media not

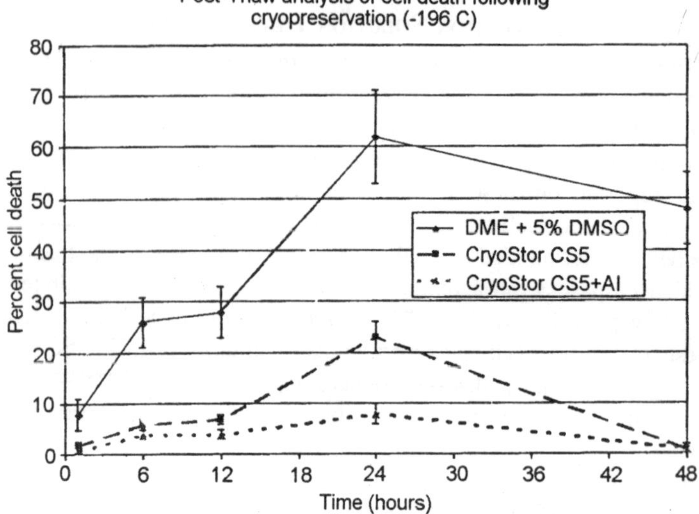

Fig. 2.1. Time course analysis of post-thaw survival of MDCK cells following cryopreservation at -196°C in cell culture plus DMSO, an extracellular-like preservation solution and in two intracellular-like solutions: CryoStor CS5 contains 5% DMSO while CryoStor CS5 + AI contains 5% DMSO plus a caspase pan-inhibitor.

only affects survival but also compromises cellular function including both the genome and proteome. Primary hepatocytes preserved in 10% DMSO effectively lose the ability to maintain normal proteomic function. In studies by Sosef et al. and Sugimachi et al., surviving hepatocytes failed to recover complete protein synthetic capabilities even fourteen days post-thaw.

The examples above reveal the dichotomy that exists in interpreting post-thaw "survival." The cause of these response differentials is not related to cryoprotectant concentration. The 24-hour post-preservation outcome obtained with human dermal fibroblasts preserved in fibroblast growth media (FGM), an extracellular-like buffered electrolyte solution, and DMSO-free CryoStor base, an intracellular-like buffered electrolyte. To each basal media DMSO was titrated to concentrations of 0.5 to 10%. Fibroblasts were cryopreserved in these media at -196°C and post-thaw survival measured. The classic "cryopreservation cap" is apparent as survival increased with DMSO levels to a putative "optimum" followed by a decline in survival at higher DMSO levels. Four significant observations may be drawn from these data. First, "*optimal survival*" is confirmed in the FBM -5% DMSO, extracellular-like media. Second, with a simple change in cryoprotectant carrier

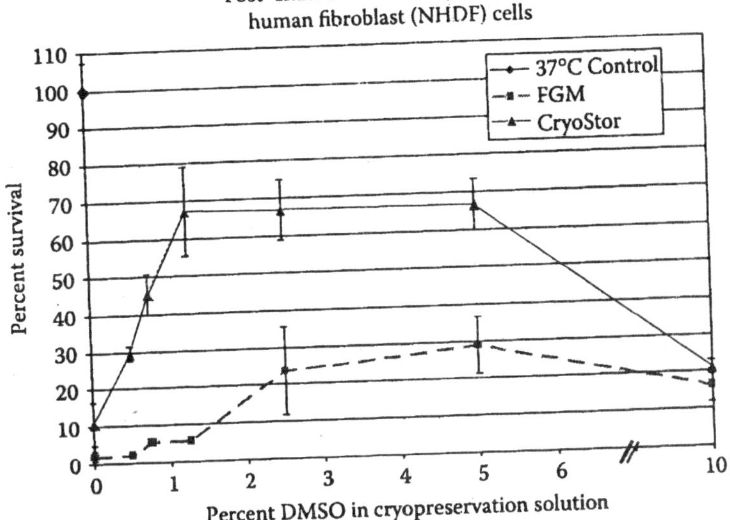

Fig. 2.2. Human dermal fibroblasts were cryopreserved in an extracellular-like (FGM) solution or an intracellular-like solution both containing 0-10% DMSO.

media from the extracellular-like formulation, post-thaw survival in CryoStor with 5% DMSO yielded an approximate doubling in survival. Third, fibroblast survival is maintained at supraoptimal levels in the Cryostor base even when DMSO levels are decreased fourfold to 1.25%. Fourth, equivalent levels of survival are obtained in FGM -5% DMSO and CryoStor +0.5% DMSO, which represents a tenfold decrease in penetrating cryoprotectant levels. This patterned response has been observed in many of the analyzed cell systems and exposes a limitation of the two-factor hypothesis.

The principal difference in the approaches to cryopreservation and the resultant outcomes dependent on carrier solution composition. Essentially, through the integration of an intracellular-type solution with an understanding of the molecular responses of the cell at low temperature, one is able to significantly improve cryopreservation outcome. Critical to an understanding of the success of these intracellular-like solutions is an in-depth knowledge of the cell death pathways activated as a result of cryopreservation-induced oxidative stresses. The processes and pathways associated with the induction of apoptosis and necrosis are reviewed along with a discussion of the current state of knowledge with regard to the extent and activation sites of these molecular-based events. While a detailed understanding of these processes remains, in its infancy, the literature base detailing this paradigm shift has rapidly expanded over the last five years.

Advances in Hypothermic Storage Solutions

The development and formulation of hypothermic storage solutions involves addressing a complex array of factors. Many of these principles were applied in the development of the first intracellular-like preservation media ViaSpan, by Southard and Belzer. Through solution design that addressed the complexity of factors influencing cell viability, as understood in the 1980s, ViaSpan proved successful at short-term cell, tissue, and organ storage. The development of this single solution was essential to the application, success, and expansion of the field of organ transplantation over the last few decades. However, current day demands for hypothermic storage of cells, tissues, and organs challenge the capabilities of ViaSpan.

While few of the new solutions have overcome the many regulatory hurdles commonplace to clinical settings, several solutions have been developed that show the promise of improved post-storage quality based on *in vitro* and *in vivo* studies. In 1995, Taylor et al. published on the development of a new generation hypothermic storage solution termed HypoThermosol. Throughout the 1990s, studies of this solution focused on its utilization as a whole-body perfusate in support of ultra profound hypothermic surgery in canine, rabbit, porcine, and nonhuman primate models. Based upon the success of these studies, utilization was extended into cell and tissue systems *in vitro* in the late 1990s. With successful translation *in vitro*, a series of studies were launched to look at the comparative differences in solution composition effects at the cellular and molecular level. Studies focusing on the effects of divalent cation dis-regulation, ion alternatives, free radical damage, protease activation, etc. have appeared and helped to guide the development of hypothermic storage solutions. Of these new directions in solution design based on cellular response processes, the control of free radical accumulation during the preservation process has proven most beneficial to date. Several recent reports by Mathew et al. have detailed the benefits of including specific free radical scavengers in hypothermic preservation solutions. These studies, built on the solution design approaches put forth by Southard et al., and later refined by Taylor et al. demonstrated that one could significantly improve hypothermic preservation efficacy as well as extend the preservation window (cold storage time). More recently, Snyder et al. demonstrated that through utilization of the hypothermic storage approach (solution) developed by Taylor et al.

and modified by Mathew et al., onc could also maintain cellular function at high levels. The resultant solution HypoThermosol-FRS was able to preserve cardiomyocytes *in vitro* for three days at 4°C while maintaining cell viability and contractile function near nonpreserved control levels.

Next Generation Cryopreservation Solution Design

With cryopreservation advancement more dependent on mitigation of the molecular-based responses to low-temperature exposure and storage, there is a growing recognition that cryopreservation solution formulation requires attention to both physical and cellular related events. The underlying principles now recognized as essential to optimization of cryopreservation processes include but are not limited to maintaining proper cold-dependent ionic ratios, control of pH at lowered temperatures, prevention of the formation of free radicals and avoidance of their detrimental effects, oncontic balance, the supply of energy substitutes, etc. In fact, the first attempt at addressing these issues was through the simple addition of cryoprotective agents such as DMSO to hypothermic storage solutions including ViaSpan. These studies showed marginal improvement in cryopreservation outcome over classical approaches. In 1998 the first report on the use of a newly designed hypothermic storage solution supplemented with DMSO was shown to markedly increase cell survival in comparison to both classical approaches as well as the use of ViaSpan + DMSO. Subsequent studies utilizing this platform solution and molecular-based approach resulted in the development of the first intracellular-like cryopreservation solution, CryoStor. This approach has led to the paradigm shift related to the control of stress response of cells during the preservation process in an effort to reduce the levels of post-thaw apoptosis and necrosis. The combination of molecular control and classical cryopreservation principles and the resultant technologies has proven to be extremely successful in improving preservation efficacy in many cellular and tissue systems.

Closing Thoughts

It is now essential that we begin to integrate the disciplines of cellular biology, molecular biology, cryobiology, etc. to effect improvement in preservation outcome. Many cell-based applications in cell therapy, regenerative and reparative medicine, biobanking, tissue engineering, and drug discovery require normal, predictable, and timely return to function of cellular systems following preservation, which is often not the case with today's standard technologies and approaches. Furthermore, with the growing body of evidence suggesting that the often utilized, elevated levels of cryoprotective agents, such as DMSO, impact and alter the cellular proteome, genome, and fragmentome — in addition to cellular structures such as the mitochondria, cell membrane, and nucleus — it is now becoming more imperative for the development of new approaches to afford successful preservation.

The first generation of cryopreservation developments focused on structural preservation of cells through the inclusion of penetrating cryoprotectants and the resultant management of ice and chemosmotic perturbations. Second-generation strategies integrated with and improved upon preservation outcome through preservation solution designs that mitigated the detrimental effects of preservation-induced oxidative stresses that directly contributed to the post-thaw launch of apoptotic and necrotic cell death cascades. The next generation of improved preservation technologies will be linked to "management" of gene-regulated, stress-dependent effects on the proteome and genome, which will likely require a near-team reassessment of the inclusion of toxic/mutanagenic agents, such as DMSO, in preservation media.

The literature base utilizing the integrated approach to understanding and developing new approaches for preservation continues to grow slowly. However, this growth is still significantly outpaced by reports dedicated to methodological improvements focusing on more precise modeling and analysis of physical parameters and volume shifts of cells undergoing preservation. While this was the case in the past and still is today, it will mostly likely be a union between the optimized structural preservation approaches and the cell-molecular-based approaches to preservation that will ultimately yield true improvements in post-preservation outcome.

3

BIOMATERIALS

The replacement of tissues or organs lost through damage or disease has provided significant clinical challenges over many years. Restoration of normal function of the tissue or organ is the ultimate goal of tissue engineering and regenerative medicine. The application of tissue engineering is not only directed at finding suitable replacement materials to restore tissue loss and function, but more recently has expanded to include application to "*biomanufacturing*" processes. The combination of the appropriate cells, with the appropriate biomaterial to produce the appropriate protein, enzyme, or cell progeny for example, is only the beginning of future possibilities in biomanufacturing.

Biomaterial is defined as "material that is compatible with living tissue — material that can be safely implanted into the human body and left there without causing an adverse reaction" (*Encarta Dictionary*, UK). This definition can be extended to include material that is compatible with living cells, *ex vivo*.

There are several categories of biomaterials including those derived from living tissue (human and animal) either with (e.g., whole organs, skin, bone, tendon grafts) or without (e.g., acellular dermal matrix, demineralized bone, collagen extracts) cellular components present, recombinant sources (e.g., bacterial-derived, mammalian cell-derived), or synthesized sources (e.g., polylactic acid and its derivatives, alloys, ceramics). Whatever the source, for successful therapeutic or manufacturing application, the material must be biocompatible. Biocompatibility is broadly defined as "the ability of a material to perform with an appropriate response in a specific application." However, consequences of such things as source, preparation or processing, preservation or stabilization, storage, handling, combination of materials, degradation, integration, and remodeling can all affect the biocompatibility of the material, which is therefore not just dependent on the properties of the material itself. All of these facets of biomaterial development must be addressed for successful therapeutic or biomanufacturing outcomes. The very nature of a biocompatible substance requires interaction with its surroundings, whether it be implanted, grafted or used *ex vivo* (e.g., manufacturing) to elicit an appropriate response. Fundamental interactions between surfaces, cells, and tissues are described by Dee, Puleo, and Bizios, and basic biological responses to biomaterials are outlined by Anderson and will not be detailed here.

Historical

Tissue transplantation has been performed for hundreds of years. There were few early successes as a consequence of the lack of understanding of immunobiology, compared to the present where it has become routine with the selection of materials to be used ever expanding and not only the number of surgical applications growing, but application of the matrices or materials to a variety of therapeutic and manufacturing needs.

In this chapter, we attempt to illustrate some of the different biomaterial strategies used to create both biologically derived and synthetic therapeutic products. In the field of tissue engineering, a biocompatible, off-the-shelf, efficacious biomaterial is the goal, which all tissue engineering strategies strive to achieve. The application of tissue engineering to manufacturing of cells and cell by-products requires cellular biocompatibility, but the complication of an *in vivo* immune response is eliminated.

Historically, the most efficacious biomaterials are autografts. Biocompatibility is not of concern, as the transfer of tissue from one part of the body to another is recognized as "self" and not foreign. These can be considered as vital tissues (living) that contain an active cellular component. The problems that exist are limitations of autograft supply. For example, in full-thickness skin for burns, if a major proportion of the body is burned, in both pediatric and elderly patients, where donor sites are limited, procuring tissue with a viable blood supply, or in repeated coronary bypass surgery where all available vessels have been utilized. The donor site itself is a second surgery with removal of the tissue for subsequent placement. This removal also results in trauma, increased morbidity, and scarring depending on the amount or depth of tissue removed. Allograft organ transplantation is routine these days with the advent of effective immunosuppressive therapy. These vital or "living" tissues require rapid transference to recipients, and preservation techniques such as simple refrigeration for short-term storage and transport. Longer-term storage would allow a delay in transferring to a recipient, which would result in an accumulation of organs to meet demand. However, the use of vital tissues, such as organs, for allogeneic transplant requires life-long immune suppression therapies to prevent rejection of the organ. There is also increased risk of disease transmission when cells are present. Organs including heart, lung, and kidney are typically used, when they are available, as life-saving treatments for patients. Complications include lifetime immunosuppressive therapy, rejection and disease transmission such as bacterial contamination, viruses, and malignancies. Early bone transplants also showed evidence of residual tumor. However, availability of donated organs is limited and emergence of diseases with late onset [e.g., Creutzfeldt-Jacob disease (CJD) and new variant (nv) CJD or "mad-cow" disease] or lack of available testing procedures for viruses (e.g., West Nile virus) increases the risk of disease transmission to recipient patients.

Replacement of various body parts (nose, ears, eyelids, etc.) has been described as early as during ancient times and could arguably be the first foray into tissue engineering. Viable allograft tissue transplants have and are still used without immunosuppression (bone, cornea, and skin). These tissues can be stored longer through cryopreservation and hence have more general supply, but have similar risks to organ transplant and, in addition, may result in alloimmunization against subsequent grafts.

There are several types of potential biomaterials that can be divided into two major categories: biological and synthetic. Combining biological and synthetic materials along with other additives such as growth factors, genes, and cells leads to a plethora of possible devices for clinical use and for biomanufacturing purposes.

Biological Approach

Several biologically derived materials have been used clinically. These include viable tissues, decellularized materials, and extracted proteins that have subsequently been synthetically derived or combined with specific growth factors.

Many aspects of effective tissue repair, replacement, or regeneration need to be considered when developing alternative materials. These include whether the material will be rejected, whether it will affect the surrounding tissue (biocompatible), the mechanical strength and durability, the long-term viability including revascularization, incorporation of the material (as part of the natural architecture of the site or conversely, encapsulated by the body keeping it away from surrounding tissue), and finally, remodeling of the material preferably into the same tissue as the implant site.

Biological approaches to tissue replacement have always attracted significant interest. This has been seen in the widespread use of donated tissue transplants including cornea, bone, heart valves, blood vessels, ligaments, tendons and cartilage for tissue replacement, and skin as a temporary dressing in the treatment of full-thickness burns. Again while having advantages such as heart valves not requiring anticoagulation, there are limitations based on the potential for disease transmission and alloimmunization. Alloimmunization results from cells within the tissue transplant potentially sensitizing the recipients to future transplants.

Nonvital or acellular matrices such as demineralized bone and acellular soft tissue grafts, such as dermis, have been developed to overcome the problems of cell sensitization and to a degree, disease transmission while maintaining the advantages of biological materials able to interface and become involved in the body's regeneration response. The effectiveness of these grafts has led to the development of machined or engineered bone grafts such as ramps, spacers, and dowels for a variety of orthopedic procedures (such as cervical, thoracolumbar, and lumbar spine interbody, fusions and pinning grafts). In addition, mineralized bone (powder and cancellous chips) and demineralized bone (cortical bone particles and fibers) have been combined with a variety of binding agents such as glycerol, gelatin (porcine-derived), calcium sulphate and carboxymethyl cellulose to produce a variety of pastes, putties, gels, and flexible sheets to markedly increase the variety of grafts available for surgical use. The addition of carriers assists in the ease of handling for the surgeon and expands the use of the donor-derived material so that many patients are able to benefit from its use. These products demonstrate that a variety of synthetics and biological (gelatin and bone) materials can be combined, producing clinically useful biomaterials.

The inherent growth factors (mainly bone morphogenic protein, BMP) within the demineralized bone matrix provide the biological signaling required to direct bone formation. More recently, the growth factor itself (BMP-2) was approved for specific indications in combination with two different carriers; namely, a bovine-derived collagen sponge, to be used in combination with titanium spinal fusion cages (FDA approved) and for general orthopedic use, with a calcium phosphate carrier. Calcium phosphate is the major mineral component of bone and is relatively inexpensive and bioinert and functions primarily as a delivery vehicle for the growth factor.

An acellular dermal matrix derived from human skin was developed that has retained the biological and physical properties of the matrix. Allograft skin has been used for decades as an effective, temporary dressing for full-thickness skin injury. Due to the presence of cellular histocompatiblity complex antigens, the host ultimately rejects allogeneic skin. The extracellular matrix of the dermis is a complex three-dimensional array of information. This information is in the form of proteins such as multiple types of collagen, each with different molecular motifs. There are also numerous proteoglycans that form an interface between collagens and cells and provide a reservoir for growth factors, cytokines, and morphogens. This biochemical information helps direct the repopulating cells to regenerate normal tissue and restore function. Several attempts have been made to produce extracellular matrix biosynthetically; however, in the absence of a complete knowledge of matrix design and without a complete and available array of all the proteins and growth factors, such attempts to date have met with limited success. Other methods have also been used to process allogeneic skin to deliver components of the dermal extracellular matrix. In one such approach, the dermal component was derived from human skin by soaking the skin in a buffered saline solution for weeks until the epidermis spontaneously released and the cells of the dermis had necrosed. In another approach human skin was repeatedly frozen and thawed to produce a devitalized human dermis for keratinocyte seeding. While showing promising *in vitro* results, no reports of effective clinical application have been published using these methods.

The core AlloDerm acellular matrix technology was a new method of freeze-drying cells and tissues without the damaging effects of ice crystals. This involved the creation and drying of the amorphous or glass phase of ice. This technology was subsequently expanded to enable the prolonged dry preservation of any tissue type. To this preservation technology was added tissue processing or tissue engineering technology in the form of cell removal (decellularization) and preservation of the remaining extracellular matrix. The result of the application of these patented technologies is to produce a structurally and biochemically intact acellular, extracellular matrix. As all cells are removed, the matrix does not elicit a specific immune response and as the matrix is preserved and not altered or damaged during processing, little if any nonspecific inflammatory response is stimulated following transplantation. Rather, because the biochemistry of the matrix is preserved, it contains the necessary information to stimulate host cell migration into the matrix resulting in revascularization and repopulation of the matrix, thereby creating a viable tissue that becomes an integrated part of the host.

Clinical use of the acellular dermal matrix technology started in 1993 in the US, first as a skin graft in the treatment of full-thickness burns and subsequently expanded to a wide range of applications. The clinical applications now extend to the following areas:

1. Burns: predominantly in the pediatric and geriatric age groups and in extensive burns for grafting of major joints and hand reconstruction
2. Other skin conditions such as Epidermolysis Bullosa (EB) where skin defects lead to extensive scarring and contracture
3. Head and neck reconstruction as a soft tissue implant
4. Reconstructive surgery, for scar revisions, correction of soft tissue deficits due to trauma, cancer surgery or congenital abnormalities, or wasting with disease
5. Interpositional graft and a facial sling (static reanimation for facial paralysis, for example, in patients with strokes)
6. Periodontal surgery to increase the amount of attached gingiva (reduced as a result of gum disease)
7. Urology as a bladder sling for urinary incontinence
8. Gynecology for pelvic floor repair — surgical intervention is usually required for pelvic floor repair for the treatment of cystocele (bladder), enterocele (intestine), and rectocele (rectum) and vault suspension for uterine prolapse
9. General surgery for repair of abdominal wall defects and hernias; fascial grafts have been used historically and can better resist infection when compared to a commonly used synthetic mesh. The major problem with the use of the fascial grafts, usually harvested from the fascia lata in the thigh, is donor site trauma, including fluid accumulation, delayed movement, and pain
10. Orthopedic surgery as a periosteal replacement membrane and for rotator cuff repair

11. Repair of damaged heart tissue resulting from aneurysm or chronic ischemia and for which no donor heart is available
12. Breast reconstruction surgery
13. Neurosurgery (off label) as a dural substitute in duraplasty. This application is classified as a medical device and has not yet been approved by the FDA
14. Chronic ulcer repair

In addition to direct soft tissue replacement, application in certain areas, for example, gingiva, dura, and bone, the acellular matrix has demonstrated a potential to remodel into tissue indistinguishable from the position of transplantation. Its original characteristics of dermal tissue are replaced by the characteristics of the local surrounding tissue.

The acellular dermal matrix has been shown to function in a manner equivalent to autograft dermis taken from a donor site on the patient. The processing of this matrix in the manner described maximizes the use of donated human skin. The matrix has the advantage of being able to be stored for two years under standard refrigeration. The other advantage of this type of tissue is that it can respond to changing physiological stresses and stimuli when implanted whereas synthetic materials cannot respond.

Another acellular matrix that has been used in clinical procedures is small intestinal submucosa (SIS) (SurgiSis, Cook Surgical). It is derived from porcine small intestine; the submucosa is found between the mucosal and muscular layers of the small intestine. SIS is extracted from the intestine in a manner that removes all cells and has been used in colon and rectal surgery, ENT/head and neck, gastroenterology, general surgery, obstetrics and gynecology, otolaryngology, plastic surgery, thoracic surgery, urology, and vascular surgery.

Collagen extracted from animal- and human-derived sources has been used for years. Type I collagen derived from bovine and porcine sources is predominantly used in the medical device industry.

Future Directions

Although very effective and with a long record of accomplishment of clinical safety, it is recognized that tissue grafts will always be limited by supply of the precious gift of tissue donation. This has led to a strong interest in the cross-discipline area of tissue engineering. The basic tenet of tissue engineering is that tissues are composed of cells and scaffolds in the presence of growth factors and cytokines. Tissue engineering in current context is usually considered the combining of cells (various types), scaffold (biological or synthetic, biocompatible, integrating) and growth factors, genes, proteins, peptides (or any other bioactive molecule) to result in the replacement, regeneration, or enhancement of a nonfunctional, damaged, or missing tissue. The result would be to recapitulate the biological and physical function of the tissue deficit with minimal or no adverse event.

In the 1950s the focus was on the use of materials that were bioinert and had minimal if any interaction with surrounding tissue. These materials, including ceramics, Teflon, silicone, metal alloys, and titanium, were and are still used to replace joints (hips), blood vessels, heart valves, and bone. Although effective, these materials are limited in their durability and/or biocompatibility. Degradation of synthetic materials can release harmful by-products locally into the surrounding tissues and systemically. Hence, mechanical hips have a finite lifespan, and mechanical heart valves require lifelong anticoagulation to prevent clotting.

Synthetic biomaterials may be divided into their material class of ceramics, metals, polymers, and composites. They may also be classified by the body response or degradation rate, hence leading to descriptors such as inert, biotolerant, bioactive, and resorbable. In these situations the body will either surround a biomaterial if it is inert or biotolerant or may create tissue which bonds closely with the biomaterial. In each application, there will be situations that can stimulate a cell response and so a biological classification of implants may provide descriptors of conductivity and inductivity, the ability to conduct tissue growth or have a specific tissue growth in a part of the body that does not normally include that particular tissue.

Biomaterials will be discussed in terms of their materials classification, their architectural modification for tissue engineering, and a comparison made with biological alternatives. A combination of biological and synthetic approaches will be considered as a means of improving the body's participation in healing or creating a specific tissue type.

Classes of Biomaterials

All materials used for implantation must satisfy the requirement of biocompatibility. As a first step, materials should not lead to systemic toxicity, mutagenicity, carcinogenicity, pyrogenicity, irritation, or sensitization. After these criteria are met, the material may be considered as a biomaterial for augmentation or other application. At the first level a biomaterial may be biotolerant, in which case the body will isolate the implanted object with a fibrous capsule to remove any influence on the body function. Inert materials may also induce a fibrous capsule due to the competition between different tissue types.

Ceramics fall into the class of inert biomaterials. They typically have a high elastic modulus and are very stiff. Some ceramics exhibit a high abrasion resistance and are thus used in articulating surfaces such as joint prosthesis. The low fracture toughness limits the applications to environments subject to compressive stresses, or the absence of tensile stresses. Examples include alumina, zirconia, and glasses. Inert polymeric materials are also common such as silicone, polyethylene, and polymethylmethacrylate. These represent the first generation of materials used in implantation.

Metals also have a high stiffness, like ceramics, but are tougher and are conventionally used for load-bearing applications or situations where a repetitive loading is experienced such as heart valves. Metals have the disadvantage in that the pH and oxygen concentration within the body do not provide immunity for the metal. Dissolution occurs at a higher level than ceramics and is referred to as corrosion when higher levels are attained. This is undesirable since specific elements can disturb the cell function and may also lead to rejection of an implant as a result of the hypersensitivity. To improve the surface of these alloys a surface coating of a ceramic such as hydroxyapatite on prosthesis or carbon on heart valves is used. Typical alloys are stainless steel, Ti, Co-Cr, and Zr based alloys. The flexibility of producing metallic alloy compositions can potentially give rise to a bioresorbable material. However, the availability of high concentrations of a metal and the influence on cell function will remain an issue. This can lead to delayed-type hypersensitivity.

From the bioactive materials that are able to create a direct bond to the tissue, there are naturally occurring materials and others that undergo a surface modification to integrate with the tissue. Hydroxyapatite is a naturally occurring material, found within bone and teeth, and can be a site for apatite precipitation and collagen deposition, thus forming a link with surrounding bone tissue. Although this material is widely used, the optimum arrangement of crystals and chemical composition is a feature that requires ongoing investigations. Bioactive glass is subjected to a modification by hydrolysis, incorporation of calcium ions followed by direct bonding to bone. Both materials are bioactive, but possess a low ability to sustain loads. Studies of calcium carbonate have also shown direct tissue bonding, which is particularly attractive due to its porous structure.

Degradable materials are the materials of choice for patients who possess the ability for reconstruction of tissue. These materials will degrade in the body, providing free space for the cells to generate tissue and provide the potential for tissue engineering. The degradation products need to be biocompatible, must be readily removed, and must not lead to an undesired cell response. Some elements have a desirable action at low concentrations, but an overdose can lead to a negative response. The composition of biomaterials and their degradation products are critical. The removal pathway of degradation products and the exposure to different body components is not well known, and therefore other side effects may only be revealed after a prolonged period. Conventional metallic alloys that participate in some movement such as lubrication have caused metallosis, but weight-bearing, structural members or electrode materials are engulfed with tissue and lead to a low release of degradation products.

Conventional metallic alloys are typically not addressed for tissue engineering, but novel alloys that slowly release elements to stimulate cell response may be developed as degradable candidates. Inorganic materials offer possibilities since they occur naturally in the body. From the class of bioceramics, degradable materials include calcium compounds such as calcium phosphates or bioglasses. Calcium phosphates can be produced to a range of cystallinities. As a noncrystalline phase, like a glass, amorphous calcium phosphate exhibits very high solubility for fast delivery of calcium and phosphate, a mechanism the body uses to supply calcium phosphate to infants through breast milk. A majority of polymeric materials have been used as degradable bone implants or controlled release applications. Polylactic acid has been developed three decades ago and has received FDA approval for biomedical application. The hydrophobicity and release of lactic acid upon degradation is not favorable for cell response.

Various approaches have been used to produce a more suitable chemistry. A higher degradation may be achieved by producing a copolymer with poly(glycolic acid), but the release of acidic components is still present and can be overcome by a thin layer of a more biocompatible polymer or the inclusion of calcium phosphate that will balance the pH upon degradation. Other synthetic biomaterials that are produced include poly(ε-caprolactone), poly(propylene fumarate), polyphosphazenes, polyphosphoesters, polyanhydrides, and poly(orthoesters). Development of new polymeric compositions is fast relative to metallic alloys and inorganic materials. Rapid testing approaches have been developed to quickly ascertain the suitability of different polymers for biomedical applications. Advances in natural macromolecules are slower due to a larger flexibility and conformations possible. Natural polymers have included collagen and polysaccharides (alginate, chitosan, and hyaluronate).

Influence of the Manufacturing Process

Processing of biomaterials has utilized existing processes and materials that are used in other industries. Additives are occasionally used with polymers to improve the processability or final properties of the product. The influence of these processing additives or performance enhancers such as UV stabilizers will be unnecessary in the body and may provoke a biological response. Equipment must thus be totally dedicated for biomaterials processing. The reaction must be well controlled in terms of reactant purity and reaction conditions. Some biomaterials are very sensitive to the surrounding chemical environment and may include elements or chemical groups into the structure or absorbed onto the surface of the product. Undesired by-products need to be removed by rinsing or heating, or an alternative method.

Scaffold Design and Production

The approach of using porous materials was proposed by Langer and Vacanti in 1993. To replace a volume of tissue/organ within the body, the use of a porous material offers a high surface area to accommodate cells. The porous material would be constructed from a bioresorbable material that would slowly degrade and provide space as the cells follow a particular lineage and generate tissue.

The design aspects of the scaffold involve a long list of desired attributes. A higher proportion of pore space is desirable to accommodate more cells within the porous material. The pore size is the next critical aspect, and previous work with porous materials for tissue augmentation has revealed the importance of pore size for tissue in-growth. This will be dependent on the cell type, and is greater than 100 microns for bone cells, for example Axons are cell extensions and require a smaller pore size for growth into and passage through the structure. Other factors that will influence the performance of a scaffold involve pore shape and pore connectivity. The connectivity between the pores is dictated by pore throat characteristics such as pore throat size, curvature from the previous pore and the number of connections with neighboring pores. Lastly, the shape of the surface may influence the cell response, and the possibilities include surface area, surface curvature, and roughness.

Fluid flow throughout the scaffold can thus be improved with pore gradients or including passages within a porous material. Pores with little connectivity may be designed to provide less fluid flow, receiving nutrients or disposing of waste product from a diffusion process to the surrounding fluid.

Scaffolds for Biomedical Function

One of the major goals in developing biomaterials is to minimize the immune response and maximize the regenerative response.

Biomaterials represent a possible alternative to repairing an organ or tissue. Synthetic biomaterials have been investigated due to the ease of manufacturing large quantities of the desired biological tissue replacement.

The scaffold may need to sustain a load, and several possibilities are available. First, the biomaterial will have a set stiffness and strength. For example, polymers typically exhibit low strengths and rigidity. Incorporation of filler with a higher stiffness such as a calcium phosphate or bioglass will increase the stiffness with a possible increase in strength. This approach would require the selection of components that have a similar degradation rate. Pore design aspects can also be used to improve the stiffness and strength. Smaller pore content will increase the stiffness, the resistance to deformation. Given a set pore volume, increasing the pore size will produce thicker pore walls, providing a better ability to sustain a load. Finally, introducing micropores into the pore walls will provide an improvement. The combination of scaffold design aspects provides an unlimited range in options, and the requirements for fluid flow, mechanical properties, etc. and need to be prioritized to provide the best scaffold performance.

Clinical applications of load-bearing synthetics are primarily in spinal arthrodesis (or fusion) procedures. Materials used in these fusion procedures have also been used for bone loss due to trauma, neoplasia, deformity, infection, and degenerative disease. Recent developments in spinal implants have improved the success rate of these fusion procedures, which have been in use for more than fifty years. A large percentage of spinal fusion procedures involve an interbody fusion between adjacent vertebral bodies. In these procedures, the lordosis is corrected using a spacer or strut that is placed between or adjacent to the vertebral bodies where disk or vertebral bone has been removed. To ensure a successful procedure, this spacer or strut must function both as a structural support and as a conduit for bone growth and fusion. Implants are currently available for virtually every interbody application, including anterior cervical discectomy and corpectomy, anterior lumbar and posterior lumbar interbody fusion, and thoracic and lumbar vertebrectomy.

There are various approaches for producing porous materials for tissue engineering. At the simplest level, one may require a sheet that exhibits porosity. Meshes or intertwined fibers, an approach used widely in the textile industry, have been used for this application. The use of fibers represents one scaffold development direction. Woven or knitted fiber meshes provide a variable pore size, but lack structural stability for clinical use. This led to the development of immobilizing the separate fibers by fiber bonding. Fiber-bonded structures provide a high porosity, but lack the required strength for load-bearing tissues. Fiber allocation initially was by a mechanical process where structures were woven or knitted, then carefully placed within a specific orientation for different layers, and the newest approach has involved a random deposition of fibers in as process called electrospinning. This produces a high pore content scaffold with control over fiber diameter as low as 50 nm, but little control of pore size.

In the 1980s the materials used for scaffolds were resorbable and predominantly amorphous, lacking specific tissue architecture and forming nonspecific interaction with cells. Many of the scaffold materials were derived from components that had been used for a long time in resorbable sutures. Hence, Dermagraft, a product used to treat chronic ulcers, was a porous scaffold of two materials known as polylactic acid and polyglycolic acid materials used in Vicryl sutures. Onto this lattice neonatal fibroblasts were cultured. The concept was that the cells would produce their own matrix while the original scaffold gradually resorbed so ultimately a tissue would be formed.

As this and similar products were tested, it became evident that it was necessary to engineer more complex tissue mechanisms such as tissue orientation and components and revascularization or angiogenesis. A cell cannot survive in general if it is more than 200 microns from a blood supply unless it is part of a special tissue such as articular cartilage or cornea. In addition, it has become increasingly recognized that the interaction between the cell and the surrounding matrix is critical and involves specific factors that interact with special receptors on the cell surface and determines the nature and function of the cell. This has led to the development of biointeractive materials that are resorbable and have information or components within the scaffold able to stimulate specific cell responses at a molecular level. These components can be growth factors, integrins, or a variety of extracellular matrix molecules or proteins such as various types of collagens, hyaluronic acid, laminin, fibrinogen, and vitronectin or heparan sulphate.

These biointeractive or "smart" materials are being used in new types of manufacturing procedures such as photolithography or creating layers of smart surfaces, which can be combined to mimic the three-dimensional structures of tissues. Another method is so-called organ printing using CAD design to sequentially deposit layers of materials to again build a three-dimensional representation of tissue. Into these constructs are embedded growth factors and other agents that can stimulate blood vessel formation and potentially specific cell behavior. Surface topography has been revealed as a key component that can direct cell movement and stimulate a particular response. Progress in the scaffold design is addressing the incorporation of nano and microtopography, and only one approach can presently provide a fine topography, a series of fine ledges on the walls of the scaffold that may direct cell movement into neighboring pores.

Smart materials are also being incorporated in new cell culturing techniques. The incorporation of bioreactor technology in combination with smart surfaces has enabled the tissue constructs to be subjected to flow and physiological stresses during cell growth and differentiation. These bioreactors are designed to mimic the natural environment of the cells, that is, within the body. These smart materials enable stem cells to be used as the seeding cell population. The concept is that stem cells, which theoretically are an indefinitely renewable cell source, can provide a sustainable source of cells for tissue regeneration, and can be directed to form the required type of cell for the tissue construct.

An example of the historical progression toward tissue transplantation and bioreactor technology is seen in the evolution of the heart valve. In 1953, the Campbell and Hufnagel mechanical heart valve was introduced. This valve had a ball in a cage configuration and required long-term anticoagulant therapy to avoid life-threatening clotting. Mechanical valves have been significantly refined over the years and remain today the most common heart valve replacement but still require anticoagulant treatment. Glutaraldehyde cross-linked porcine or pig valves were introduced in 1975. These tissue valves did not require anticoagulant treatment, but because of the cross-linking and other factors, remain an inert nonliving valve with a relatively short lifespan due to calcification and tissue deterioration.

Cryopreserved or frozen human heart valves have had increased use from 1985, mainly in the pediatric age group. These valves are in relatively short supply and can cause alloimmunization or sensitization of the recipient to other transplants. More recently, early results of a tissue-engineered heart valve were published by Hoerstrup and co-workers. The group fabricated a valve from poly-4-hydroxybutrate coated polyglycolic acid. The construct was then seeded with endothelial and smooth muscle cells and placed in a bioreactor under physiological pressures and flow for 14 days. The valve was implanted into sheep for up to 20 weeks, and normal valve function was demonstrated.

Incorporation of bioactive factors (growth factors, cytokines, morphogens, and even genes that have been altered themselves so that they can be released slowly, rapidly, on demand or in a specific sequence) into a smart material as part of its design can specifically instruct the cells placed on the material what to do and when to do it. For example, Mahoney and Saltzman developed a polymeric material with soluble and insoluble components so that the nerve growth factor embedded within each component was released at different times. Cells adhered onto this surface were subsequently implanted and the release of nerve growth factor was detected for 21 days.

A more basic approach has been to inject marrow-derived stem cells directly into the damaged heart in the hope that they seed in the damaged tissue and repopulate, differentiate, and replace the damaged tissue. The efficacy of this approach has recently been disputed. Less than 2% of injected cells can ultimately be detected. Tissue regeneration is a constant battle between scar formation and normal tissue regeneration. If the balance is tipped toward scar formation, then the injected cells will only see the damaged tissue and will respond to this environment accordingly. If the cells see undamaged tissue or a compatible matrix with appropriate embedded information, then they are more likely to respond by regenerating the tissue.

Due to the recent advances in stem cell technology, stem-cell-based and non stem-cell-based tissue engineering is being addressed on several fronts including blood vessels, liver, bladder, meniscus, bone, pancreas, cartilage (ear, nose, joints), oral mucosa, nervous tissue, salivary gland, cornea, skeletal muscle, heart valves, trachea, dentin, skin, intestine, ureter, heart muscle, kidney, urethra, and potentially others. Bone marrow contains stem cells that have the capacity to develop into numerous tissues some of which are listed above as well as blood cells (angioblastic cells). Another aspect of the use of smart surfaces is to expand certain types of cells outside the body to be used as a separate source of cells.

Traditionally, red cells, white cells, and blood plasma are taken from blood donors and then separated into these components. The number of suitable donors is diminishing with increased health risks to some populations. This is evidenced by the emergence of several recent cases of CJD in the UK following distribution of blood from a single CJD positive donor (not detected at the time of donation and subsequently presented with the disease) to a number of recipients who subsequently contracted the disease. Blood donations have reached crisis point in the UK, but the reach is widespread as potential donors may have lived in the UK for greater than 6 months: "The deferral implemented by the Red Cross is the result of guidance issued by the U.S. Food and Drug Administration (FDA). The guidance calls for the indefinite deferral of all blood donors who have spent a cumulative time of six months or more in the United Kingdom between the years 1980 and 1996. The United Kingdom includes England, Northern Ireland, Scotland, Wales, Isle of Man, or the Channel Islands. The deferral is intended to reduce the *theoretical* risk of transmitting new variant Creutzfeldt Jakob disease (nvCJD) through blood transfusion". If the different types of blood cells could be produced in bioreactors from stem cells that have been fully characterized, then the disease risk would be negligible. Applications of tissue engineering to these problems have enormous potential benefits to health care around the world.

Combining cell and tissue engineering technologies translates into significant prospects for the future for these disciplines. Combining biomaterials, scaffolds, stem cells, and progenitor cells can create biointeractive products to address significant medical needs, such as major tissue deficits and organ damage or failure.

Several tissue engineering initiatives worldwide are bringing together expertise to address specific problems. For example, the Pittsburgh Tissue Engineering Initiative lists its target areas as oral/ bone, cardiovascular, stem cell, nerve, muscle, and skin tissue engineering. Under these areas, its directed research focus is listed as human developmental biology as applied to tissue engineering, polymer-based advanced biomaterials program, tissue imaging and bioinformatics as applied to tissue structure and function, gene therapy as applied to tissue engineering, artificial organs and tissue engineering, and clinical trials for tissue engineering technologies.

Clearly, several disciplines need to be brought together in a synergistic manner to address clinically relevant needs. This will potentially produce a number of therapeutic outcomes and research tools during the development of these therapeutic products. Other such initiatives include: Rice University Bioengineering Program implementation of a cross-disciplinary research training for their bioengineers capitalizing on recent advances in molecular and cell biology in order to translate scientific aspects of biotechnology into new cost-effective products and processes; and the Georgia Tech/Emory Center (GTEC) for the Engineering of Living Tissues. GTEC's mission is to be the leader in the development of critical core technologies and an educated workforce that will enable the development of tissue engineering, revolutionize the medical implant industry, and in the process, help confront the transplantation crisis.

Stem cells, smart surfaces, smart materials, bioreactors, and scientific ingenuity hold great promise to make significant advances in tissue transplantation, tissue regeneration, and biomanufacturing in the not too distant future.

4

TISSUE PRESERVATION

The purpose of this chapter is to provide an overview of the "*state of the art*" for biopreservation of tissues ranging from simple clumps of cells to complex tissues containing multiple cell types. Preservation of specific tissue types including ovaries, corneas, vascular grafts, articular cartilage, heart valves, skin, and pancreatic islets have been the subject of recent reviews. The use of human allogeneic tissues, such as skin, heart valves, and blood vessels in medicine has become normal practice. However, while clinical demand for tissues continues to grow, supply of these valuable human resources has become a limiting factor. As a result, the development of living engineered constructs has become an important new field of biomedical science and biopreservation of donor tissues and manufactured product is generally recognized as an important issue without which many market applications may never be fully achieved.

The differences between simple cell suspensions and structured, multicellular tissues with respect to their responses to cooling, warming, and dehydration clearly impact their requirements for biopreservation. These differences have previously been described in detail with respect to the response of structured tissues to freezing and thawing. We regard biopreservation to be a crucial enabling technology for the progression from preclinical and translational clinical research on cellular tissue products for regenerative medicine and transplantation. Tissue biopreservation is also needed for samples to be used for various research and toxicology test purposes. The need and advantages of tissue biopreservation are widely recognized and well documented. There are several approaches to biopreservation, the optimum choice of which is dictated by the nature and complexity of the tissue and the required length of storage. Obviously, tissues such as bone and tendon that are banked successfully without a viable cell component (often referred to as nonliving tissues) are far more robust in withstanding the stresses of preservation than "living" tissues that invariably contain cells that must retain viability for maintenance of tissue functions. Maintenance of structural and functional integrity of living tissues is demanding and will be the principal focus of this chapter.

Short-term preservation of tissues and organs that cannot yet be successfully cryopreserved because they sustain too much injury at deep subzero temperatures can be achieved using either normothermic organ culture, in the case of corneas, or more commonly hypothermic storage at temperatures a few degrees above the freezing point. Normothermic organ culture is usually limited to corneal preservation for periods of up to a month or more. In contrast, hypothermic storage is commonly employed for many types of tissue for transport between tissue donation sites, processing laboratories and end users, and short-term storage. These approaches to short-term biopreservation are dealt with in other chapters and will be covered only in outline here for completeness.

With the exception of normothermic organ culture, all approaches to biopreservation aim to stabilize biological tissues by inhibiting metabolism and significantly retarding the chemical and biochemical processes responsible for degradation during *ex vivo* storage. Long-term preservation calls for much lower temperatures than short-term hypothermic storage and requires the tissue to withstand the rigors of heat and mass transfer during protocols designed to optimize cooling and warming in the presence of cryoprotective agents (CPAs). As we will explain below, ice formation in structured tissues during cryopreservation is the single most critical factor that severely restricts the extent to which tissues can survive cryopreservation procedures involving freezing and thawing. In recent years, this major problem has been effectively circumvented for several tissues by using ice-free cryopreservation techniques based upon *vitrification*.

Long-term preservation in the presence of ice is achieved by coupling temperature reduction with cellular dehydration. In principle, stabilization by dehydration without concomitant cooling can be achieved for long-term storage at ambient temperatures. However, the application of this approach to mammalian systems is in its infancy and will be addressed at the end of this chapter as a prospective development for the future.

Obtaining adequate and reproducible results for cryopreservation of most tissues requires an understanding of the major variables involved in tissue processing and subsequent preservation. Optimization of these variables must be derived for each tissue by experimentation guided by an understanding of the chemistry, biophysics, and toxicology of cryobiology. Before discussing the factors affecting tissue quality in detail, it is necessary to consider the meaning of "viability" with respect to tissue function *in vivo*. Viability may simply be defined as the ability of preserved tissues to perform their normal functions upon return to physiological conditions. Many means of assessing cell viability within tissues have been described including amino acid uptake, protein synthesis, contractility, dye uptake, ribonucleic acid synthesis, and 2-deoxyglucose phosphorylation. The assay(s) used to determine viability should give clear indications that the cells are alive and, preferably, should report on activities important for long-term tissue functions.

Multicellular tissues are considerably more complex than single cells, both structurally and functionally, and this is reflected in their requirements for cryopreservation. Some cell systems, such as platelets and sperm, may be subject to thermal or cold shock upon cooling without freezing. In general, tissues are not known to be sensitive to cold shock. However, due to concerns that CPAs, such as DMSO, may increase tissue sensitivity to cold shock or result in cytotoxicity, CPAs are usually added to tissues after an initial cooling to 4°C prior to further cooling. Frozen tissues have extensive extracellular and interstitial ice formation following use of tissue bank cryopreservation procedures. Such frozen tissues may, in some cases, have excellent cell viability. In other cases, as discussed later, viability may be very poor or cell viability can be good but cells in the tissue may no longer operate as a functional unit. It is usually not possible to detect where the ice was present after thawing by routine histopathology methods. Cryosubstitution techniques, which reveal where ice was present in tissues, have, however, demonstrated significant extracellular tissue matrix distortion and damage. The extent of freezing damage depends upon the amount of free water in the system and the ability of that water to crystallize during cooling.

Other factors, in addition to ice formation, have biological consequences during biopreservation: the inhibitory effects of low temperatures on chemical and physical processes and, perhaps most important, the physiochemical effects of rising solute concentrations as the volume of liquid water decreases during crystallization. The latter process results in cell volume decreases, pH changes, and the risk of solute precipitation. There have been several hypotheses on mechanisms of freezing- induced injury based upon such factors.

Two main approaches to tissue cryopreservation are currently in use or development. The first involves traditional freezing methods, based upon the cell preservation methods developed shortly after the Second World War, versus approaches involving ice-free vitrification. Both approaches involve the application of cryobiological principles. Cryobiology may be defined as the study of the effects of temperatures lower than normal physiologic ranges upon biologic systems. Simply cooling cells or tissues with spontaneous ice nucleation and crystal growth results in dead, nonfunctional materials.

Little advance was made in the field of cryobiology, with respect to significant post-freeze cell survival, until 1949 when Polge et al. reported the cryoprotective properties of glycerol. Shortly thereafter, Lovelock and Bishop discovered that dimethyl sulfoxidc (DMSO) was also an effective cryoprotectant. Since the discovery of these CPAs, the field of cryopreservation has flourished and many other cryoprotectants have been identified. Before considering cryopreservation of tissues further, we will review some of the principles of short-term preservation in the absence of freezing.

Short-term Tissue Preservation

This section deals with the issues relating to the selection and design of solutions for hypothermic preservation and tissue transport. Most tissues are transported on ice for short periods of time before being processed for an application or cryopreserved for long-term storage. More specifically, the composition of the buffer medium used to nurture the tissue during the preservation protocol is very important but often overlooked on the assumption that conventional salt buffers such as Ringer's lactate and Krebs' solution, or regular tissue culture medium will be adequate. There are very good reasons that these types of solutions may be suboptimal for cell preservation in tissues at reduced, hypothermic temperatures.

Since control of the component cell and tissue environments is of fundamental importance for the outcome of preservation, it is worth considering some of the basic principles of solution design in relation to low-temperature storage. In essence, two basic types of solution are considered and often referred to as "*extracellular-type*" or "*intracellular-type*" solutions to reflect their basic ionic composition.

Over the years we have encountered some confusion and inconsistencies in the use of these terms, hence definitions of these terms are footnoted.

Some more sophisticated hypothermic blood substitution solutions such as Hypothermosol, Unisol, and KPS1, the formulations also contain an oncotic agent in the form of a high-molecular-weight colloid such as Dextran or hydroxyethyl starch. These solutions are perfused through the vascular bed of an individual organ, or even the whole body, at a pressure sufficient (typically 25–60 mm Hg) to achieve uniform tissue distribution. To balance this applied hydrostatic pressure and prevent interstitial edema, an oncotic agent such as albumin or synthetic macromolecular colloids (e.g., Dextran or hydroxyethyl starch) is incorporated into the perfusate. These perfusates may have an "*intracellular*" or "*extracellular*" complement of ions depending upon the temperature of perfusion preservation.

Methods of short-term hypothermic preservation are neither standardized nor optimized for various tissues and organs. Currently the formulation of solutions employed differs depending upon the type of tissue or organ and whether the excised organs are stored statically on ice or mechanically perfused. Historically, a variety of preservation solutions have been developed and used for organs, but their application for tissues has been generally neglected. Most tissues are still transported in cell culture media and the fallacy of assuming that physiological culture media is acceptable for low-temperature storage is illustrated below. While there are undisputed industry standards for certain organs and applications, the concept of a "*universal*" preservation solution for all tissues and organs has still to be realized in practice. In general, the solutions adopted for abdominal visceral organs (kidney, liver, and pancreas) have not proved optimal for thoracic organs (heart and lungs) and vice versa. In contrast, a new approach to "universal" tissue preservation solutions has been developed for bloodless surgery using *hypothermic blood substitution* (HBS) to protect the whole body during profound hypothermic circulatory arrest (clinical suspended animation, or "*corporoplegia*" — literally meaning body paralysis).

In recent years we have used this approach based upon the "Unisol" concept, in which two new solutions (a "*maintenance*" and a "*purge*") formulated for separate roles in the procedure have been tested. The principal solution is a hyperkalemic, "*intracellular-type*" solution designed to "*maintain*" cellular integrity during hypothermic exposure at the nadir temperature (< 10°C). The companion solution is an "*extracellular-type*" purge solution designed to interface between blood and the *maintenance* solution during both cooling and warming. This novel approach to clinical suspended animation has been established in several large animal models and more recently explored for resuscitation after traumatic hemorrhagic shock in preclinical models relevant to both civilian and military applications. Most recently the efficacy of hypothermic blood substitution with the hybrid Unisol solutions for whole-body protection has been demonstrated in a porcine model of *uncontrolled lethal hemorrhage* (ULH).

In considering the efficacy of a solution for universal tissue preservation there is no better test than to expose all the tissues of the body to cold ischemia. Moreover, protection of those tissues most exquisitely sensitive to an ischemia/hypoxia insult, namely the heart and brain, provides the greatest challenge. Current interests in the development of hypothermic arrest techniques to facilitate resuscitation of hemorrhagic shock victims in trauma medicine has parenthetically provided an opportunity to examine the efficacy of new hypothermic blood substitution solutions for universal tissue preservation. In accordance with earlier baseline models, these studies validate the unequivocal benefit of profound hypothermia combined with specially designed blood replacement fluids for the protection of heart, brain, and the other major organs during several hours of cardiac arrest and ensuing ischemia. Moreover, the preservation of higher cognitive functions in these animals subjected to hypothermic arrest further corroborates previous reports of the high level of neuroprotection provided by these hypothermic blood substitute solutions.

The general *in vivo* tissue-preservation qualities of this hypothermic blood substitution technique are clearly demonstrated by the consistent resuscitation of exsanguinating animals after traumatic hypovolemic shock. More specifically, biochemical profiles of the surviving animals showed that, apart from a transient elevation in liver enzymes that normalized within the first post-op week, there were no metabolic signs of organ dysfunction. This is consistent with previously published observations in a canine model of clinical suspended animation in which specific markers of heart, brain, and muscle injury (creatine kinase isozymes) all showed transient increases in the immediate post-op period and returned to normal baseline levels within the first post-op week.

In conclusion, the demonstrated efficacy of these synthetic, acellular hypothermic blood substitute solutions for protection of all the tissues and organs in the body during clinical suspended animation justifies their consideration for multiple organ harvesting from cadaveric and heart- beating donors. Furthermore, these observations support the findings of parallel studies for longer-term hypothermic storage of a variety of cell types derived from vascular tissues and kidney in which the Unisol-UHK *maintenance* solution has provided excellent cell survival compared with a variety of commonly employed solutions. This provides further evidence for the protective properties of hypothermic blood substitutes such as *Hypothermosol* and *Unisol* solutions used for global tissue preservation during whole body perfusion in which the microvasculature of the heart and brain are especially vulnerable to ischemic injury. Moreover, the application of solution-design for clinical suspended animation under conditions of ultraprofound hypothermia places the Hypothermosol and Unisol solutions in a unique category as universal preservation media for all tissues in the body. In contrast, all other preservation media, including the most widely used commercial solutions such as UW-ViaSpan are established for specific organs, or groups of organs, e.g., UW for abdominal organs and Celsior or Cardiosol for thoracic organs.

Hypothermic Storage of Tissues

Hypothermic storage of whole organs flushed or perfused with a preservation solution is common practice in clinical transplantation. This procedure leaves the vascular endothelial cells in direct contact with the preservation medium during the cold ischemic period. The effect of storage conditions on the integrity of vascular endothelium is therefore of crucial importance for the quality of preservation of intact organs. Moreover, it has been established in recent years that the long-term patency of vascular grafts used in reconstructive surgery may be significantly affected by cold storage. We have reviewed the effects of hypothermia upon vascular function elsewhere. The importance of these effects is illustrated by some experiments we conducted to compare the microscopic changes in tissue morphology when excised blood vessels were immersed and transported in Unisol-UHK hypothermic preservation solution compared with Dulbecco's Minimum Essential Medium (DMEM), which is a common culture medium used to incubate and transport tissues *ex vivo*. The total cold ischemia time in these experiments was relatively short, 34.5 ± 9.5 minutes for samples transported in DMEM, and 37 ± 4 minutes for samples in Unisol-UHK. Tissue immersed and transported in DMEM on ice exhibited microscopic changes within the tunica intima and tunica media. The intima was intact, but there was extreme vacuolization of the underlying basal lamina causing, in turn, extrusion of the endothelial cells into the lumen and giving a "rounding-up" appearance. Apart from this vacuolization, the endothelial cells had a near normal appearance. The smooth muscle cells (SM) had a somewhat shrunken appearance with irregular contours. The tunica adventitia was essentially normal.

In contrast, jugular veins transported in Unisol demonstrated little if any histological changes compared with the DMEM group. The tunica intima was intact with little evidence of vacuolization of the underlying basement membrane. The smooth muscle cells (SM) did not appear shrunken and were in a normal, horizontal orientation.

These preliminary observations have been extended recently in a comprehensive study of interactive determinants that impact the preservation of autologous vascular grafts. A multifactorial analysis of variance was used in the design and analysis of a study to evaluate the interaction of solution composition with time and temperature of storage. In summary, these *in vitro* studies that have examined both the function and structure of hypothermically stored blood vessels have clearly shown that optimum preservation of vascular grafts is achieved by selection of the type of storage medium in relation to time and temperature. Synthetic preservation solutions such as Unisol, designed specifically to inhibit detrimental cellular changes that ensue from ischemia and hypoxia, are clearly better than culture medium or blood and saline, which are commonly used clinically.

Basic Rationale for the Design of Hypothermic Solutions

The scientific rationale for the choice of components selected in the design of Unisol is largely the same as that described previously for other hypothermic solutions and will be summarized here. The strategic design of solutions used for organ preservation have differed depending upon their ultimate use, either as a flush solution for static storage of the organ, or as a perfusate for continuous, or intermittent, perfusion of the organ. Taking a unique approach, Unisol has been formulated with a view to developing a universal baseline solution that may be used for both hypothermic static storage of tissues and organs, and also for machine perfusion preservation. By design Unisol contains components that help to (1) minimize cell and tissue swelling, (2) maintain appropriate ionic balance, (3) prevent a state of acidosis, (4) remove or prevent the formation of free radicals, and (5) provide substrates for the regeneration of high-energy compounds and stimulate recovery upon rewarming and reperfusion. Parenthetically, these are regarded as the minimum essential characteristics for what can be regarded as baseline formulations. Additional classes of compounds can be considered as additives to these baseline solutions to fine-tune the cytoprotective properties. The scientific basis for new strategies to counteract cold ischemic injury by modifying storage solutions and perfusates is still emerging. Most of these strategies focus on combating oxidative stress and cold or hypoxia-induced apoptosis. Some insights into the cellular and molecular mechanisms of cold-storage injury have recently been reviewed by Rauen and DeGroot. While these mechanisms have been elucidated using various experimental models, potential strategies to counteract their effects have yet to be demonstrated in clinical practice.

The strategic design of Unisol included combining the main characteristics of effective hypothermic solutions with attention toward selection of multifunction components. By carefully selecting components that possess multiple properties, the intrinsic protective properties of these hybrid solutions are maximized.

A fundamental biophysical property of the Unisol design is to provide the optimum concentration of ions and colloids to maintain ionic and osmotic balance within the organ or body tissues during hypothermia. In particular, an effective impermeant anion is included to partially replace chloride in the extracellular space and prevent osmotic cell swelling (i.e., to balance the fixed ions inside cells that are responsible for the oncotic pressure leading to osmotic cell swelling and eventual lysis during ischemia and hypothermia). A number of anions including citrate, glycerophosphate, gluconate, and lactobionate, or the anionic forms of aminosulphonic acids such as HEPES could be suitable candidates. Lactobionate (FW = 358) was used exclusively as the principal impermeant in many solutions developed in recent years; these include ViaSpan, Hypothermosol, Celsior, Cardiosol, and Churchill's solution, among others. Lactobionate is also known to be a strong chelator of calcium and iron and may, therefore, contribute to minimizing cell injury due to calcium influx and free radical formation. However, for organ perfusion Belzer and Southard recommended against using lactobionate in a perfusion solution, especially for kidneys. Instead, gluconate was selected and shown to be an important component of Belzer's machine perfusion solution. As a hybrid solution, Unisol has effectively incorporated both impermeants and uses gluconate (70 mM) as the predominant impermeant plus lactobionate (30 mM) for its additional cytoprotective properties.

The osmoticum of Unisol is supplemented by the inclusion of sucrose and mannitol, the latter of which also possesses properties as a hydroxyl radical scavenger and reduces vascular resistance by inducing a prostaglandin-mediated vasodilatation that may be of additional benefit.

A macromolecular oncotic agent is an important component of a perfusate to help maintain oncotic pressure equivalent to that of blood plasma. Any oncotic agent that is sufficiently large to prevent or restrict escape from the circulation by traversing the fenestration of the capillary bed may be considered. Examples of acceptable colloidal osmotic agents include:

1. blood plasma expanders such as human serum albumin
2. hetastarch or hydroxyethyl starch (HES) — an artificial colloid derived from a waxy starch and composed almost entirely of amylopectin with hydroxyethyl ether groups introduced into the alpha (1–4) linked glucose units
3. Haemaccel (Hoechst) — a gelatin polypeptide
4. pluronic F108 (BASF — a nonionic detergent copolymer of polyoxyethylene and polyoxypropylene
5. polyethylene glycol
6. polysaccharide polymers of D-glucose such as the dextrans

For a variety of reasons, dextran-40 (average mol wt = 40,000 daltons) was selected as the preferred colloid of choice for oncotic support to balance the hydrostatic pressure of perfusion and help prevent interstitial edema. It has long been known that dextran can improve the efficiency of the removal of erythrocytes from the microvasculature of cooled organs by inhibiting red cell clumping and by increasing intravascular osmotic pressure and reducing vascular resistance. These attributes of dextran may be particularly important during washout, both *in vivo* and *ex vivo*. Dextran is widely used clinically as a plasma expander and is readily and rapidly excreted by the kidneys. There is ample recent evidence that dextran-40 is also an effective and well tolerated colloid in modern cold-storage solutions for organ preservation. Moreover, in 1996, a Swiss clinical study verified that dextran-40 safely replaces HES in the UW solution for the purpose of human kidney preservation for transplantation.

Retention of the colloid in the vascular space is an important consideration for achieving optimal oncotic support. Any dextran that permeates into the interstitial space during the hypothermic procedure will be readily eluted upon return to physiological conditions. Another advantage of the use of dextran is that the viscosity of the solution will not be as high as with some other colloids such as HES.

The ionic balance, notably the Na^+/K^+ and Ca^{2+}/Mg^{2+} ratios, of Unisol has been adjusted to restrict passive diffusional exchange at low temperatures when ionic pumps are inactivated. The concentration of monovalent cations Na^+ and K^+ are approximately equimolar to restrict their passive transmembrane exchange. Due to documented concerns regarding the very high potassium levels in commercial organ preservation solutions, including ViaSpan and EuroCollins, the potassium concentration in Unisol is lower by comparison, but sufficiently elevated to fulfill the requirements of an "*intracellular-type*" preservation medium.

In the area of cardioplegia and myocardial preservation there is good evidence for improved survival using elevated concentrations of magnesium and very low, but not zero, calcium to avoid the putative calcium paradox. Some glucose is included in these hypothermic solutions as a substrate, but the concentration is low to prevent exogenous overload during hypothermia, which may potentiate lactate production and intracellular acidosis by anaerobic glycolysis.

Acidosis is a particular hazard during hypothermia and attention has been given to the inclusion of a pH buffer that will be effective under nonphysiological conditions that prevail at low temperatures. HEPES was selected as one of the most widely used biocompatible aminosulphonic acid buffers that have been shown to possess superior buffering capacity at low temperatures, and have been included as a major component of other hypothermic tissue preservation media. Synthetic zwitterionic buffers such as HEPES also contribute to osmotic support in the extracellular compartment by virtue of their molecular size (HEPES = 238 daltons).

Adenosine is a multifaceted molecule and is included in the hypothermic preservation solutions not only as an essential substrate for the regeneration of ATP during rewarming, but also as a vasoactive component to facilitate efficient vascular flushing by vasodilatation. Glutathione is included as an important cellular antioxidant and hydroxyl radical scavenger, as well as a cofactor for glutathione peroxidase, which enables metabolism of lipid peroxides and hydrogen peroxide.

Unisol, by design, is a base vehicle solution to which any combination of pharmacological additives might be added for optimization of the preservation of a particular tissue or organ. Moreover, there is the potential benefit of a wide variety of pharmacological and biochemical agents that may be selected from the following categories:

1. Cytoprotective agents and membrane stabilizers
2. Energy-producing substrates and nucleotide precursors
3. Calcium channel blockers
4. Oxygen-derived free radical scavengers/antioxidants
5. Apoptosis inhibitors
6. Vasoactive agents
7. Trophic factors
8. Molecules for oxygen delivery

Tissue Screening, Preparation, and Antibiotic Sterilization

If tissues are destined for transplantation, it is normal practice for them to go through an extensive donor screening, microbial testing before and after processing to check for potential contaminants, and incubation in antibiotic formulations to hopefully kill low levels of microbial contaminants that are below the resolution of the microbial sampling protocols employed. In the USA the United States Food and Drug Administration (FDA) regulates tissue-engineered products, and the allograft tissue community has its own regulatory body, the American Association of Tissue Banks (AATB). The AATB has established standards that provide minimum performance requirements for all aspects of tissue banking activity. These guidelines include requirements for donor suitability as well as the handling of transplantable human tissue. Their intent is to assure allograft tissue recipients disease- and contaminant-free implants, and to help ensure the optimum clinical performance of transplanted cells and tissues.

In contrast, tissue-engineered product regulations are in various stages of development by the FDA. One of the issues from a regulatory perspective has been that many tissue-engineered products combine features of drugs, devices, and/or biological products, so the normal division of work structure at the FDA has had to be modified. The Center for Biologics Evaluation and Research (CBER), the Center for Devices and Radiological Health (CDRH), and the Center for Drug Evaluation and Research (CDER) have established intercenter agreements to clarify product jurisdictional issues within the FDA. The FDA has also established an InterCenter Tissue Engineering Work Group to address scientific and regulatory issues and members of this group have been very active at large. Two other special interest groups are presently involved in the development of standards for tissue-engineered products, the American Society for Testing and Materials (ASTM) and the Tissue Engineering Special Interest Group of the Society for Biomaterials.

Recently the FDA published final rules establishing donor eligibility criteria for donors of human cells, tissues, and cellular- and tissue-based products (HCT/Ps) to help prevent the transmission of communicable disease when these products are transplanted. This new rule is part of the agency's plan to regulate tissues and related products with a comprehensive, risk-based approach. The new rule on donor eligibility pertains to donors of traditional tissues such as musculo-skeletal, skin, and eye tissues that have been required to be screened and tested for HIV, hepatitis B virus (HBV), and hepatitis C virus (HCV) since 1993. Under this new rule, reproductive tissue (semen, ova, and embryos), hematopoietic stem cells derived from cord blood and peripheral blood sources (circulating blood sources as opposed to bone marrow), cellular therapies, and other innovative products are also regulated. In addition to including a broader range of tissues and cells, the new rule extends the scope of protection against additional communicable diseases that can be transmitted through transplanted tissues and cells. The new regulation adds requirements to screen for human transmissible spongiform encephalopathies, including Creutzfeldt-Jakob disease (CJD), and to screen and test for syphilis. Screening and testing for still other relevant communicable disease agents [human T-lymphotropic virus (HTLV)] would be required for viable cells and tissue rich in leukocytes such as semen and hematopoietic stem cells. The new rule also provides a framework for identifying emerging diseases that may pose risks to recipients of transplanted HCT/Ps and for which appropriate screening measures or testing are available. Thus, this new regulation gives the FDA the flexibility to rapidly address new disease threats as they appear, providing substantial new protections for patients receiving tissue transplants. Examples of such diseases include West Nile virus and Severe Acute Respiratory Syndrome (SARS). The rule also contains requirements related to record keeping, quarantine, storage, and labeling of the HCT/Ps, all important to the prevention of disease transmission. The final rule became effective on May 25, 2005.

Regardless of whether the transplantable material is an allograft or an engineered product, recipient safety must be ensured through administration of strict donor screening criteria along with stringent quality control measures that encompass the entire tissue preparation protocol. In contrast to most inanimate medical products, which are sterilized to eliminate bacterial, viral, and fungal contaminants, it is not possible to terminally sterilize products that contain living cells. It is necessary, therefore, to ensure the safety of these products by stringent control of the living component source.

In engineered products, just as with allogeneic tissues and organs, procedures for the screening of donors and handling of materials must be strictly performed. These procedures are presently in the process of being defined and may vary depending upon the type of cell source employed. There is no need for modification from existing clinical practice in the autologous situation, in which cells or tissues are removed from a patient and transplanted back into the same patient in a single surgical procedure. If the autologous cells or tissues are banked, transported, or processed with other donor cells or tissues, however, there then exists opportunities for product adulteration and the introduction of transmissible disease. When this is the case, good manufacturing practices (GMPs) and good tissue practices (GTPs) should be implemented and it becomes necessary to screen for infectious agents. For example, in the case of the CarticelTM process, in which biopsies of healthy cartilage are used as a source of chondrocytes, the biopsies are minced, washed, and cultured with cell culture medium containing antibiotics. However, years ago Brittberg et al. found that presence of antibiotics (50 μg/ml gentamicin sulphate and 2 μg/ml amphotericin B) in the culture medium may prevent contaminant detection; therefore the culture medium should be changed to an antibiotic-free formulation prior to testing for bacterial contamination and extensive washing of the biological material may be required to remove inhibitory antibiotics to allow their proliferation and subsequent detection.

Utilization of allogeneic donors is associated with greater risk than an autologous donation because of the risk of infectious disease transmission from the donor to the recipient. Donor screening similar to that outlined in the following section on allogeneic tissue grafts should be performed, along with the implementation of product screening and/or quarantine procedures. The allogeneic cells may then be (1) used with no modifications after expansion *in vitro*, (2) genetically manipulated, or (3) turned into continuously proliferating cell lines. As an additional safety measure, specimens of each donor tissue should be archived for future pathogen screening.

There is a similar concern about unidentified diseases in the use of xenogeneic cell, tissue, and organ sources. The range of infectious diseases potentially transmissible by xenogeneic cells is unknown because infectious agents that produce little or no effects in animals may have severe consequences in human patients. The FDA has published draft recommendations on infectious disease issues in xenotransplantation. Donor screening issues again play a significant role in the prevention of infection. With xenogeneic materials, the pedigree and health status of environmentally isolated herds or colonies of animals to be used as donors become critically important.

All tissues should be delivered to patients with the highest possible assurance that they are free of pathogens. The two most effective and common methods for the prevention of infectious agent transmission are thorough donor screening and adherence to sterile techniques during harvesting, transport, and processing. The first step in confirming a potential donor in most parts of the world is to obtain permission for organ and/or tissue donation from the donor's legal next of kin. Regulations in some countries and states may differ. Once permission has been obtained, the donor must be screened to minimize the potential for transfer of infectious or neoplastic disease. In most organ donors, the donor has been declared brain dead and organ functions are supported by extracorporeal life support. This is to ensure maintenance of tissue viability while permission for donation is obtained and donor screening and infectious disease testing is performed. The AATB standards require donor tissue be tested for blood-borne infectious disease prior to acceptance for transplantation, so tissues are placed in quarantine until the test results are complete.

The most commonly employed viable tissue allografts involving antiseptic treatments are tricuspid heart valves (aortic and pulmonary valves) and blood vessels. Other tissues, such as skin where viability is not considered a major issue and, less frequently, transplanted tissues, are subject to antiseptic treatment methods similar to those used for heart valves. Donor tissues for transplantation are obtained aseptically in an operating room or, alternatively, at autopsy in a clean fashion. The donor is usually prepared in a manner similar to preparing the incision site of a patient for surgery. For example, skin donors are usually shaved and the areas of skin to be removed are scrubbed with an iodine-based wash, rinsed with isopropyl alcohol, and again painted with iodine. Kirklin and Barratt-Boyes have presented surgical techniques for the recovery of hearts for valve procurement in an autopsy setting.

In order to provide a sterile allograft for transplantation, identification and elimination of any potential contaminants are required. The antiseptic treatment stage of tissue processing begins once the tissues have been prepared and dissected employing aseptic technique. AATB standards dictate that "processing shall include an antibiotic disinfection period followed by rinsing, packaging, and cryopreservation" and that "disinfection of cardiovascular tissue shall be accomplished via a time-specific antibiotic incubation." Following immersion in the antibiotic solution of choice, the tissues are incubated, while immersed, at either 4°C or 37°C for up to 24 h (temperature and time being variable between tissue banks). Following incubation the tissues are packaged aseptically for tissue storage by cryopreservation.

Many different antibiotic mixtures for treatment of tissues for transplantation have been employed. Heart valve indicators of effectiveness, documented in many studies over the years, include cellular viability, host ingrowth rate, disinfection efficiency, and valve survival rates. Various formulae using various combinations of penicillin, gentamicin, kanamycin, axlocillin, metronidazole, flucloxacillin, streptomycin, ticarcillin, methicillin, chloramphenicol, colistimethate, neomycin, erythromycin, and nystatin have been tried for heart valve treatment. Skin is usually treated with gentamicin or combinations of penicillin and streptomycin. Csonge et al. reported the best results with combinations of ceftazidime, ampicillin, and amphotericin. Currently, a modified version of the antibiotic treatment regimen recommended by Strickett et al., in which cefoxitin, lincomycin, polymyxin B, and vancomycin are added to sterile-filtered RPMI 1640 tissue culture medium, is being commonly used in the USA to disinfect allograft heart valves. Various nutrient media have also been used, including modified Hank's solution, TCM 199, Eagle's MEM, and RPMI 1640 in conjunction with various antibiotic "cocktails." While many different antibiotics in various tissue culture media have been employed, all are employed at relative low doses with varying incubation times at either 4°C or 37°C. Care should be taken to optimize the antibiotic concentrations and conditions to minimize loss of tissue viability and function while maximizing antimicrobial effectiveness.

The antibiotic solutions developed for the antiseptic treatment of heart valves were originally formulated to ensure sterility after weeks of storage at 4°C. However, Angell et al. showed that antibiotics are more effective at physiological temperatures (~37°C) than at refrigerator temperatures. The simple protocol of collection of heart valves within 24 h of death combined with low-dose antibiotic treatment has been reported to be sufficient to produce pathogen-free valves in >95% of cases. In fact, there is little evidence that antibiotics even provide bactericidal action at 4°C since most antibiotics function through interference with temperature-dependent processes of nucleic acid synthesis or the bacterial cell wall. It is possible that the effectiveness of some of the 4°C antibiotic treatment protocols can be credited to antibiotic binding during low-temperature incubation, and that upon subsequent rewarming the antibiotic action is actually activated. Nevertheless, and regardless of the mechanism, cardiovascular tissue programs usually advocate the use of antibiotics, and in some cases residual antibiotics in the tissue may actually reduce the risk of subsequent graft infection. These tissue grafts are often preferred for implantation in infected patient sites.

Besides the issues of microbial effectiveness, there is also the issue of cytotoxic effects of antibiotics upon cells and tissues. Cram et al. provided evidence that reduction in antibiotic concentrations improves viability of refrigerated stored skin. There have been many reports of antibiotic effects on heart valve cell viability, alterations in cell morphology, and inhibition of cell ingrowth. In particular, amphotericin B may destabilize mammalian cell membranes during cryopreservation by altering the mobility of the cholesterol in the lipid phase of the plasma membrane or alterations in osmoregulation. Alternatively, amphotericin B is supplied as a colloidal suspension that has been dispersed by the detergent deoxycholate. This detergent may directly alter the membrane permeability properties of mammalian cells and such changes may render the cells less resistant to the osmotic stresses of freezing and thawing. There have been few reports on alternative antifungal agents; however, Schmehl et al. assessed the water-soluble fungicide flucytosine as an alternative to amphotericin B in combination with imipenem, a wide spectrum β-lactam, but this fungicide is not being used clinically for tissue processing. Elimination of amphotericin B from the antibiotic regimen used to sterilize grafts emphasizes the fact that post-procurement treatment cannot be relied upon to guarantee recipient safety. It further highlights the importance of thorough donor screening. Permission for autopsy and obtaining pertinent medical history, including detection of symptoms associated with systemic mycoses or infective endocarditis, are paramount to exclusion of fungal contaminants originating from the donor graft. Strict sterile technique during recovery, transport at 4°C, and cold, sterile processing are additional measures to prevent fungal proliferation. A sterile specimen is preferred because evidence of fungus may be masked by an overgrowth of competitive bacteria. Coprocessed specimens are usually tested before and after antibiotic treatment, just prior to packaging for cryopreservation. Antibiotics cannot be expected to unfailingly disinfect every allograft. The issue of how to assure sterility of tissue allografts while maintaining cell viability and tissue functions has no effective solution in sight.

CRYOPRESERVATION

It is important to emphasize at the outset that successful cryopreservation of tissues is not a simple matter of extrapolating the well-established principles of cell cryopreservation to more complex tissues. The reason is that tissues are much more than the aggregate sum of their component cells. They invariably comprise a variety of cell types intimately associated with basement membranes, an extracellular matrix, and often a vascular supply such that the structure of the integrated tissue demands special considerations in its response to cryopreservation conditions for its successful preservation. These differences are manifest in additional mechanisms of injury, identified some years ago, that must be circumvented for successful preservation. Ultimately these differences in their response to freezing between individual cells and tissues are principally due to extracellular ice formation.

A variety of factors are known to influence cell survival during cryopreservation, but the role of the vehicle solution for the CPAs is often overlooked. It is generally assumed that simple salt buffers or conventional culture media used to nurture cells at physiological temperatures will also provide a suitable medium for exposure at low temperatures. In a manner similar to our earlier discussion of optimum control of the cells' environment during hypothermic short-term storage, cryopreservation also demands consideration of the chemical composition of the buffer medium used as a vehicle for the CPAs as well as the temperature to which the cells are exposed. It has been a common practice in tissue banking to use tissue culture media as the base solution for preservation media. However, for the reasons outlined above, tissue culture media, which are designed to maintain cellular function at normal physiological temperatures, are inappropriate for optimum preservation at reduced temperatures and we have long advocated the use of intracellular-type solutions as more appropriate vehicle solutions for CPAs. Maintaining the ionic and hydraulic balance within tissues during cold exposure can be better controlled in media designed to physically restrict these temperature-induced imbalances and can be applied equally to the choice of vehicle solution for adding and removing CPAs in a cryopreservation protocol. Moreover, the nature of the vehicle solution used to expose cells and tissues to cryoprotectants at low temperatures has been shown to impact the outcome of cryopreservation, and has recently become the focus of additional research aimed at optimization and attenuation of the so-called cryopreservation cap.

Table 4.1. Major cryopreservation variables

Variable
Freezing-compatible pH buffers
Vehicle solution selection (may vary with cryoprotectant selection)
Apoptosis inhibitors (may be required to get long-term post-thaw cell survival for some cells)
Cryoprotectant selection (optima may vary with vehicle solution selected)
Cooling rate
Storage temperature
Warming rate
Cryoprotectant addition/elution conditions (number of steps, temperature)

In a study of factors that influence the survival of vascular smooth muscle and endothelial cells it was discovered that the choice of carrier solution significantly impacted the optimum survival of the cells. Moreover, the survival varied with the nature of the CPA and the cell type suggesting that nature of the vehicle solution should be included as one of the variables that must be optimized for a given system. Our aim is to use the approach of a hybrid universal formulation in an attempt to nullify the wide differences in available solution choices. Baust et al. have corroborated this approach and shown that an intracellular-type solution, *Hypothermosol*, provides a significantly better vehicle solution for CPAs than a range of extracellular-type media in other cell systems.

Another common practice in tissue banking is to employ serum of animal origin in the cryopreservation formulation. Serum-free procedures have been reported for a variety of tissues and mammalian sources of serum can be removed providing that cryopreservation conditions are subsequently reoptimized.

Synthetic Cryoprotectants

Historically, serendipity has been largely responsible for most discoveries of cryoprotectants. Cryoprotectant selection for cryo-

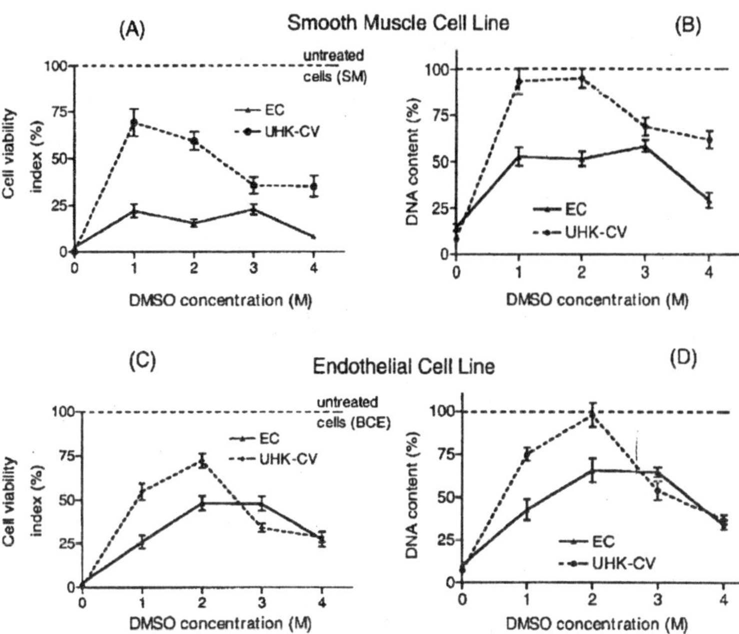

Fig. 4.1. Cell survival (viability: A and C and DNA content; B and D) after freezing and thawing with varying concentrations of DMSO in either Unisol-CV (UHK-CV) or EuroCollins.

preservation in general is usually restricted to those that have conferred cryoprotection in a variety of biological systems (dextrans, DMSO, ethylene glycol, glycerol, hydroxyethyl starch, polyvinyl-pyrrolidone, sucrose, and trehalose). Combinations of two cryoprotectants may result in additive or synergistic enhancement of cell survival. Comparison of chemicals with cryoprotectant properties has revealed no common structural features. These chemicals are usually divided into two classes: (1) intracellular cryoprotectants with low molecular weights that penetrate and permeate cells and (2) extracellular cryoprotectants with relatively high molecular weights (greater than or equal to sucrose) that neither penetrate nor permeate cells.

Intracellular cryoprotectants, such as glycerol and DMSO at concentrations from 0.5 to 3.0 molar, are effective in minimizing cell damage in slowly frozen biological systems. Extracellular cryoprotective agents such as polyvinylpyrrolidone or hydroxyethyl starch are often more effective at protecting biological systems cooled at higher rates. Such agents are often large macromolecules that affect the properties of the solution to a greater extent than would be expected from their osmotic pressure. These nonpermeating cryoprotective agents are thought to have direct protective effects on the cell membrane. This protection may be due to the oncotic forces (colloidal osmotic pressure) exerted by large molecules and alterations in the activity of the unfrozen water caused by hydrogen bonding to water molecules. Although cryoprotective agents also reduce the amount of extracellular ice at each subzero temperature with a resultant increase in the volume of the unfrozen fraction, it is not known if fewer ice crystals are responsible for any of the reduction in cell damage. The latter function of cryoprotective agents may also relate to their role in reducing membrane fusion during cryopreservation. The pharmacologic effects of cryoprotective agents such as DMSO and glycerol were reviewed by Shlafer. According to Mazur, cryoprotectants protect slowly frozen cells by one or more of the following mechanisms: suppression of salt concentrations, reduction of cell shrinkage at a given temperature, and reduction in the fraction of the solution frozen at a given temperature.

Natural Cryoprotectants

Through millions of years of evolution, nature has produced several families of proteins that may have benefits during cryopreservation, which help animals and plants survive in cold climates. These proteins are known collectively as antifreeze proteins (AFPs). AFPs have the ability to modify ice structure, the fluid properties of solutions, and the response of organisms to harsh environments. The antifreeze molecules are diverse in structure and, to date, four main types have been characterized. The first to be discovered and best characterized are the antifreeze peptides and glycoproteins (AFPs) found in Antarctic fish and northern cod species. The natural AFPs found in polar fish and certain terrestrial insects are believed to adsorb to ice by lattice-matching or by dipolar interactions along certain axes of forming ice nuclei. By default, when temperature is lowered sufficiently, growth occurs preferentially in the *c*-axis direction (perpendicular to the basal plane) in a series of steps. This abnormal growth mode produces long ice needles, or spicules, that are much more destructive to cells and tissues than normal ice. Regardless, these molecules confer a survival advantage upon certain animals. These observations led to the hypothesis that naturally occurring antifreeze molecules might be improved upon by synthesis of molecules that will either bind to other ice nuclei domains or upon stable ice crystals.

Conflicting results have been obtained by scientists following up on the proposal of Knight and Duman that many of the problems associated with ice formation during cryopreservation might be limited by the addition of naturally occurring AFP. However, the studies of Hansen et al. demonstrated that AFP Type I inhibited ice recrystallization in the extracellular milieu of cells, but increased ice crystal growth associated with the cells, and resulted in AFP concentration-dependent cell losses compared to untreated control cultures. A major focus of our research for the past six years has involved the identification of synthetic ice blockers (SIBs) that may combine with certain naturally occurring antifreeze compounds and cryoprotectants to minimize ice damage during freezing or risk of ice formation during vitrification. The best SIBs we have identified to date are 1,3-cyclohexanediol and 1,4-cyclohexanediol (patent pending).

In recent years, some of the challenges for cryopreservation of living tissues have been more fully characterized and new approaches are under development for circumventing the problems that have thus far limited the extrapolation of established principles and techniques for cells to more complex tissues and organs. A synopsis of the state of the art of biopreservation of living tissues utilizing cryopreservation methods follows.

Cryopreservation by Controlled Freezing

Successful biopreservation by freezing is dependent on the optimization of several major factors. Advances in the field of cryopreservation had been modest until Polge et al. discovered the cryoprotective properties of glycerol. Subsequent research by Lovelock and Bishop showed that dimethyl sulfoxide was also a cryoprotectant (CPA). The use of cryoprotectants during freezing and thawing of biological materials has become established and many other cryoprotectants have been identified. When cryoprotectants are used in extremely high concentrations, ice formation can be eliminated during cooling to and warming from cryogenic temperatures. Under these conditions the solution and tissue become vitrified; this is discussed further in a later section. In addition to cryoprotectant selection several major variables must be considered in development of cryopreservation methods.

A wide variety of isolated cells in suspension can be preserved using conventional cryopreservation methods involving freezing. In such methods the cells are concentrated and vitrify in ice-free channels between regions of extracellular ice. In general terms, each cell type has a freezing "window" in which the change in temperature with time provides for optimal cell survival. This proposed "window" is narrow at high temperatures and becomes increasingly wider as the temperature decreases, suggesting that deviation from a given cooling rate at high temperatures may be more critical to cell survival than deviations at low temperatures. Studies on the survival of various mammalian cell types, frozen in glycerol or DMSO, both as single cell suspensions and in tissues, frozen at a variety of rates suggest that optimal survival occurs at a cooling rate somewhere between 0.3°C and 10°C per minute.

The cell viability of cardiovascular tissues including veins, arteries, and heart valves can be preserved by a number of cryopreservation freezing techniques. However, smooth muscle and endothelial functions are usually impaired to varying degrees depending upon species, tissue type, and preservation methods employed. Clinically the fundamental issue with cryopreserved cardiovascular tissues, regardless of whether they are cryopreserved by freezing or vitrification, is that they are allogeneic and there is an inevitable immune response unless immunosuppressive therapy is employed. We have compared cryopreservation techniques in syngeneic and allogeneic rat models and concluded that the changes observed in allogeneic heart valves are primarily due to immunological incompatibility of the graft and recipient, as previously suggested in rats and human infants, not cryopreservation method. The immune response is not clinically significant for the majority of cryopreserved allogeneic heart valves; however, cryopreserved small-diameter vascular allografts typically have only short-term patency and this is probably a consequence of immunogenicity. There has been considerable discussion of whether or not cell viability is needed for allograft heart valves. The conclusion appears to be that viability correlates with "*minimally traumatized*" tissue with the result that most allograft heart valves employed today retain viable cells at the moment of implantation.

In marked contrast to cardiovascular tissue, studies using a variety of animal articular cartilage models and human cartilage biopsies have revealed no more than 20% chondrocyte viability following conventional cryopreservation procedures employing either DMSO or glycerol as cryoprotectants. Ohlendorf et al. used a bovine articular cartilage, osteochondral plug model to develop a clinical cryopreservation protocol. This protocol employed slow-rate cooling and 8% DMSO as the cryoprotectant. They observed loss of viability in all chondrocytes except those in the most superficial layer at the articular surface. Muldrew et al. previously investigated chondrocyte survival in a similar sheep model. These researchers observed cells surviving post-cryopreservation close to the articular surface and deep at the bone/cartilage interface. The middle layer was devoid of viable cells. More recently, Muldrew et al. demonstrated improved results using a step-cooling cryopreservation protocol, but cell survival post-transplantation was poor and again there was significant loss of cells in the mid-portion of the graft. The reason for lack of cell survival deeper than the superficial layers of articular cartilage is most likely multifactorial and related principally to heat and mass transfer considerations. Surface cells freeze and thaw more rapidly than cells located deep within the matrix. This phenomenon could result in a greater opportunity for ice to form, both within cells and in the extracellular matrix, deeper within the articular cartilage. Furthermore, typically employed concentrations of DMSO (8–20%) may not penetrate adequately to limit intracellular ice formation. Recent data from Jomha et al. demonstrated that increasing DMSO concentrations to 6 M can result in higher overall cell survival (40%) after cryopreservation. These observations suggest that use of higher DMSO concentrations results in better penetration of the DMSO into the cartilage.

We are aware that other factors, in addition to ice formation, may have biological consequences during freezing procedures. Two of these factors are the inhibitory effects of low temperatures on chemical and physical processes, and, perhaps more importantly, the physio-chemical effects of rising solute concentrations as the volume of liquid water decreases during crystallization. This latter process results in a decrease in cell volume and the risk of solute precipitation. Several hypotheses have been published on mechanisms of freezing-induced injury based upon such factors, but our own experiences with mammalian tissues concur with others that the principal disadvantage of conventional cryopreservation revolve primarily around ice formation.

Cryopreservation by Avoidance of Ice Formation—Vitrification

It is now generally accepted that extracellular ice formation presents a major hazard for biopreservation by freezing of multicellular tissues. This has led to a major focus during the last decade on the development of low-temperature preservation techniques that avoid ice crystallization and *ipso facto* circumvent the associated problems. The evidence for the damaging role of ice in tissue cryopreservation has been previously reported.

Prevention of freezing by vitrification means that the water in a tissue remains unfrozen in a noncrystalline state during cooling.

Vitrification is the solidification of a liquid without crystallization. As cooling proceeds, however, the molecular motions in the liquid permeating the tissue decrease. Eventually, an "*arrested liquid*" state known as a *glass* is achieved. It is this conversion of a liquid into a glass that is called vitrification (derived from *vitri*, the Greek word for glass). A glass is a liquid that is too cold or viscous to flow. A vitrified liquid is essentially a liquid in molecular stasis. Vitrification does not have any of the biologically damaging effects associated with freezing because no appreciable degradation occurs over time in living matter trapped within a vitreous matrix. Vitrification has been shown to provide effective preservation for a number of cells, including monocytes, ova, and early embryos and pancreatic islets. Vitrification is potentially applicable to all biological systems.

Vitrification preservation procedures are very similar to those employed for freezing tissues. Generally speaking, the cryoprotectants are added in stepwise or gradient manner on ice. In some cases, due to risks of toxicity, lower temperatures may be employed for the final higher CPA concentrations. The cooling rates employed are typically as fast as can be achieved for the tissue in question to temperatures around -100°C and then more slowly to the final vapor phase nitrogen storage temperature between -135°C and -160°C. Warming is performed in a similar manner, slowly to -100°C and then rapidly to 0°C. Rapid cooling and warming in our laboratory is usually performed by immersion in either chilled alcohol (-100°C) or warm water baths (4°C or 37°C).

Microwave warming has been attempted but has never been successful using conventional devices due to the uneven warming of specimens and problems with thermal runaway, which results in heat-denatured tissues. In 1990, Ruggera and Fahy reported success in warming test solutions at rates of up to about 200°C/min using a novel technology based on electromagnetic techniques (essentially microwave heating). Unfortunately, unpublished results indicate that this method is also problematic due to the uneven warming of specimens and problems associated with thermal runaway. Others have taken a systematic approach to develop a dielectric heating device to achieve uniform and high rates of temperature change. This has been achieved in some preliminary model systems but application of this warming technology with survival of cells has yet to be reported.

Vitrification and freezing (water crystallization) are not mutually exclusive processes; the crystalline phase and vitreous phase often coexist within a system. In fact part of the system vitrifies during conventional cryopreservation involving controlled freezing of cells. This occurs because during freezing the concentration of solutes in the unfrozen phase increases progressively until the point is reached when the residual solution is sufficiently concentrated to vitrify in the presence of ice. Conventional cryopreservation techniques by freezing are optimized by designing protocols that avoid intracellular freezing. Under these cooling conditions the cell contents actually vitrify due to the combined processes of dehydration, cooling, and the promotion of vitrification by intracellular macromolecules. Phase diagrams have proved to be a useful tool in understanding the physicochemical relationship between temperature, concentration, and change of phase. In particular, *supplemented* phase diagrams that combine nonequilibrium data on conventional equilibrium phase diagrams serve to depict the important transitions inherent in cooling and warming aqueous solutions of cryoprotective solutes.

The term vitrification is generally used to refer to a process in which the objective is to vitrify the whole system from the outset such that any ice formation (intracellular and extracellular) is avoided. In cryopreservation by vitrification dehydration occurs by chemical substitution alone, while in cryopreservation by freezing dehydration occurs by both osmotic dehydration and chemical substitution. In the former case the cells appear normal, while in the latter case the cells appear shrunken.

Stability of the vitreous state is critical for the retention of vitrified tissue integrity and viability. Comprehensive studies of vitreous stability for a variety of potentially important cryoprotective mixtures have been made. Glass stability of vitrified blood vessel samples stored in vapor phase liquid nitrogen storage with retention of smooth muscle function has been demonstrated up to 4 months of storage. The stability of glasses formed from aqueous solutions of 1,2-propanediol are much greater, for the same water contents, than for all other solutions of commonly used cryoprotectants including glycerol, dimethyl sulfoxide (DMSO), and ethylene glycol.

Unfortunately, solutions of polyalcoholic cryoprotectants (CPAs) such as propanediol and butanediol that show the most promise in terms of cooling rates and concentrations necessary for vitrification, also required unrealistically high heating rates to avoid devitrification during rewarming. Moreover, due principally to isomeric impurities that form a hydrate at reduced temperatures, 2,3-butanediol has proved to have an unanticipated biological toxicity at concentrations below that necessary for vitrification.

Advances in biostabilization require process development for optimization of chemical and thermal treatments to achieve maximal survival and stability. At this time the consensus opinion is that viable tissues such as blood vessels, corneas, and cartilage that have proven refractory to cryopreservation by conventional freezing methods, despite decades of intense research by many investigators, can only be successfully preserved if steps are taken to prevent or control the ice that forms during cooling and warming. In contrast, other tissues in which the cells do not function in an organized manner or in which the extracellular matrix water is not highly organized are well preserved by traditional cryopreservation by freezing methods (i.e., heart valves and skin). Our laboratory has developed a cryopreservation approach using vitrification, which thus far has demonstrated >80% preservation of smooth muscle cell viability and function in cardiovascular grafts and similar levels of chondrocyte survival in articular cartilage. In addition to *in vitro* studies of cardiovascular tissues, transplant studies have been performed that demonstrate normal *in vivo* behavior of vitrified cardiovascular and cartilaginous tissues. Most recently, this technology has been successfully applied to tissue-engineered blood vessels and encapsulated cells described below.

Avoidance of ice by vitrification has been achieved by cooling highly concentrated solutions (typically >50% w/w) that become sufficiently viscous at low temperatures to suppress crystallization rates. The original formulation and method was licensed from the American Red Cross, where it was developed for organ preservation. However, even though rabbit kidneys were successfully vitrified, they could not be rewarmed with significant retention of function. Viability was lost due to ice formation during the rewarming process. The rewarming of vitrified materials requires careful selection of heating rates sufficient to prevent significant thermal cracking, devitrification, and recrystallization during heating. The use of carefully designed warming protocols is necessary to maximize product viability and structural integrity. Vitrified materials, which may contain appreciable thermal stresses developed during cooling, may require an initial slow warming step to relieve residual thermal stresses. Dwell times in heating profiles above the glass transition should be brief to minimize the potential for devitrification and recrystallization phenomena. Rapid warming through these temperature regimes generally minimizes prominent effects of any ice crystal damage. It is presently not possible to rewarm organs rapidly enough due to their high volume relative to the volume of tissues. Development of optimum vitrification solutions requires selection of compounds with glass-forming tendencies and tolerable levels of cytotoxicity at the concentrations required to achieve vitrification. Due to the high total solute concentration within the solution, stepwise protocols are commonly employed at low temperatures for both the addition and removal of cryoprotectants to limit excessive cell volume excursions and lower the risk of cytotoxicity.

Despite developments to devise solutions that would vitrify at practically attainable cooling rates for sizeable biological tissues, the corresponding critical warming rate necessary to avoid devitrification remains a critical challenge. Conceptually, elevated pressures, electromagnetic heating, the use of naturally occurring antifreeze molecules, and synthetic ice blockers have been proposed as means to tackle the problem.

Vitrification Versus Freezing

We are often asked how we make the decision to use a vitrification approach rather than a freezing method for a particular tissue. The answer is usually a combination of method efficacy, based upon our experience and the literature with respect to the specific tissue, and ease of use or cost.

An excellent example of efficacy is conventional cryopreservation of articular cartilage by means of freezing, which typically results in death of 80–100% of the chondrocytes plus extracellular matrix damage due to ice formation. These detrimental effects are major obstacles preventing successful clinical use of osteochondral allografts and commercial success of tissue-engineered cartilage constructs. Cryosubstitution studies of frozen and vitrified articular cartilage plugs revealed negligible ice in vitrified specimens and extensive ice formation throughout frozen specimens. Transplantation studies in rabbits demonstrated that vitrified cartilage performance was not significantly different to fresh untreated cartilage. In contrast, frozen cartilage performance was significantly different when compared to either fresh or vitrified cartilage. These studies combine to demonstrate that the vitrification process results in ice-free preservation of rabbit articular cartilage plugs and that about 85% of cellular metabolic activity is retained following rewarming. In contrast, frozen tissues contained ice within the cells and the matrix, with the exception of the articular surface, where some viable cells were observed. In our experience vitrification has been superior to freezing for rabbit cartilage plugs, porcine cartilage plugs, and human biopsy specimens. That does not mean that effective freezing methods can't be developed, just that no one has come close to date.

In cardiovascular tissues the decision to use a vitrification method over freezing is not as clear cut. Both approaches to cryopreservation result in high cell viability, but in our experience smooth muscle and endothelial functions are better preserved by vitrification than freezing. At this time we hypothesize that this is due to prevention of extracellular ice damage to tissue matrix. We fully agree with criticism that we have compared vitrification with freezing methods that may be improved upon; however, we would also point out that both the vitrification and the freezing methods that we have employed may be improved by further research. We have primarily employed freezing protocols derived by extensive research and in clinical practice for allografts in the United States.

In the case of cardiovascular allografts the use of vitrification techniques may have cost benefits because there is no need for control-rate freezers. However, regardless of whether vitrification or freezing is employed, the tissue is still allogeneic with respect to the potential recipient and there is no reason to anticipate that vitrification will reduce immunogenicity. The research that we have performed on vitrification of cardiovascular tissues was intended for tissue-engineered cellular constructs; however, when the work was initiated there were no constructs available so we employed autologous and allogeneic tissue models. There are still no tissue-engineered cellular constructs approved for human use, but there are several well-established experimental models. We have compared the published vitrification and freezing procedures for two experimental vascular graft models based upon collagen or polyglycolic acid matrixes combined with smooth muscle cells. The results reflect our earlier results with rabbit jugular vein segments: cell viability was well preserved by both freezing and vitrification methods. However, in the graft model capable of developing detectable contractile forces in response to various drugs, the polyglycolic acid construct, smooth muscle function was significantly better preserved by vitrification than freezing.

In many small tissue structures (such as tissue organoids, cell aggregates, or encapsulated cells) it is anticipated that optimized freezing and vitrification procedures will provide similar levels of cell viability and tissue functions. We are finding that cryopreservation by freezing and vitrification methods may be equally effective in preservation of small pieces of tissue. The question is often which method is easiest and most consistent. We recently performed cryopreservation studies on rat embryo metanephroi (embryonic kidneys). One potential solution to xenotransplantation immunological complications is the transplantation of embryonic kidneys whose blood supply is not yet fully developed.

The metanephroi (MN) may be less immunogenic in comparison to their adult counterparts, at least in part due to the fact that post transplantation their vascular supply is derived from the host. It has been shown that MN from E15 Lewis rat embryos that are transplanted into the omentum of adult C57Bl/6J mice receiving costimulatory blockade undergo growth and differentiation. Also, the E28 pig MN growth and development occurs post-transplantation across an allogeneic or highly disparate xenogeneic barrier with costimulatory blockage. For such a therapy to be commercially viable, long-term storage of embryonic kidneys is crucial. In these studies the effects of controlled- rate freezing and ice-free vitrification on MN viability were investigated. Metanephroi were isolated from 15-day (E15) timed pregnant Lewis rats and either (1) control-rate frozen at -0.3°C/min in a DMSO formulation or (2) vitrified in VS55. The MN were then stored at -135°C for 48 h. After storage the MN were rewarmed, placed in culture media, and their viability was assessed using the alamar Blue assay and histology (light microscopy, TEM, and cryosubstitution). There were no statistical differences in embryonic kidney metabolic activity of either of the cryopreserved MN groups relative to the untreated control group.

Cryosubstitution demonstrated the presence of significant ice formation during controlled-rate freezing. This was confirmed by TEM, where vacuolation of the cytoplasm of control-rate frozen MN was observed. In contrast, no ice was observed in vitrified MN and there was very little cytoplasmic disruption. However, vitrified MN showed mitochondrial and nuclear injury suggestive of CPA cytotoxicity. This injury was not observed in frozen MN, nor have we observed similar changes in other vitrified tissues. The effects of vitrification solution formulation, concentration, exposure time, and loading steps on embryonic kidney viability need to be evaluated in future studies.

Microencapsulated Cells as Pseudo-Tissues

Another example is in the development of cryopreservation protocols for microencapsulated cells. Microcapsules are particularly prone to cryodamage using freezing by cryopreservation methods. Since 1991 studies of microcapsule cryopreservation by freezing employing a variety of cell types have accumulated. Excellent cell viability was obtained in many cases, but capsule integrity is another issue.

Viability is excellent following preservation in alginate microcapsules. Algae-derived polysaccharides, such as agarose and alginate, are a novel class of nonpermeating (with respect to the cell) cryoprotectants. These polysaccharides had no cryoprotective abilities when used alone, but resulted in enhanced viability when mixed with known penetrating cryoprotectants (such as DMSO). Ice formation in slow-rate, DMSO-protected frozen microcapsules containing insulin-secreting βTC3 cells was demonstrated using cryosubstitution at -90°C and fixation, a method that permits visualization of ice. In this same study vitrification resulted in freedom from ice. Vitrified insulin-secreting βTC3 cells had significantly better viability (metabolic activity) and function (insulin release) than frozen insulin-secreting βTC3 cells.

Very little investigation of cryopreservation variables has been performed for microencapsulated cells. Most studies have used low-rate cooling with 5–20% DMSO and storage in liquid nitrogen. Single studies have compared cryopreservation with and without nucleation control, duration of DMSO incubation prior to cryopreservation, and

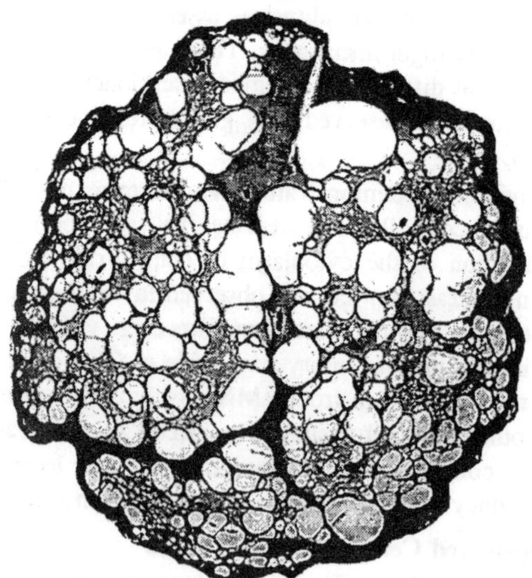

Fig. 4.2. Morphology of frozen and vitrified pancreatic substitute beads.

cooling rates with several cryoprotectant formulations. The outcome of the DMSO incubation study indicated that a 5-h incubation was required for optimum cell survival. This was curious since we have found that DMSO equilibrates in alginate capsules in 2–3 min. Nucleation was required for optimum cell survival as we would anticipate for an effective freezing cryopreservation method for either isolated cells or tissues. The most in-depth study was performed by Heng et al. In this study rapid cooling cryopreservation protocols with high DMSO concentrations (3.5 M, 25% v/v) resulted in low post-thaw cell viability (<10%), which did not improve with higher concentrations (4.5 M, 32% v/v) and longer exposure to DMSO, even though the majority of microcapsules (60–80%) remained intact. Subsequent investigations of slow cooling with a range of DMSO and EG concentrations resulted in a much higher post-thaw cell viability (80–85%), with ~60% of the microcapsules remaining intact when DMSO was used at a concentration of 2.8 M (20% v/v) and EG at a concentration of 2.7 M (15% v/v).

The presence of 0.25 M sucrose significantly improved post-thaw cell viability upon slow cooling with 2.8 M (20% v/v) DMSO, although it had no effect on microcapsule integrity. Multistep exposure and removal of sucrose did not significantly improve either post-thaw cell viability or microcapsule integrity, compared to a single-step protocol. Ficoll 20% (w/v) also did not significantly improve post-thaw cell viability and microcapsule integrity. There have been two reports on vitrification of microencapsulated cells, including our paper on microencapsulated insulin producing TC3 cells. Vitrified encapsulated cells demonstrated no significant difference between fresh and vitrified specimens and no ice was observed in the vitrified specimens. Kuleshova et al. investigated vitrification of encapsulated hepatocytes employing 40% ethylene glycol and 0.6 M sucrose. They employed a modification of vitrification approaches employed for embryos and oocytes that employs straws for handling and storage combined with a capsule made from Ter-polymer and collagen as previously described. These studies combine to demonstrate that both freezing and vitrification procedures for cryopreservation of microencapsulated cells are feasible.

However, as the scale of the tissue increases, vitrification procedures excel and levels of tissue functions and/or cell survival previously not achieved with freezing procedures are achieved. Conceptually, cryopreservation of tissues by vitrification offers several important advantages compared with procedures that allow or require ice formation. Complete vitrification eliminates concerns for the known damaging effects of intra- and extracellular ice crystallization. Furthermore, tissues cryopreserved by vitrification are exposed to less concentrated solutions of cryoprotectants for shorter time periods. For example, Rall has calculated that for embryos, during a typical cryopreservation protocol involving slow freezing to –40°C or –70°C, the cells are exposed to cryoprotectant concentrations of 21.5 and 37.6 osmolal respectively.

In contrast, cells dehydrated in vitrification solutions are exposed for much shorter periods to <18 osmolal solution, although the temperature of exposure is higher. Finally, unlike conventional cryopreservation procedures that employ freezing, vitrification does not require controlled cooling and warming at optimum rates. A principal benefit of vitrification is the elimination of requisite studies to determine optimal cooling rates for tissues with multiple cell types. Successful vitrification requires that the thermal processing be rapid enough to transition regions of maximal ice crystal nucleation and growth that occur above the glass transition temperature of the solution. Thus, it is only necessary to cool solutions at rates in which a negligible fraction of the solution forms ice (typically < 0.2%). Vitrified materials have a similar rate requirement during heating, when samples are rewarmed for subsequent use, to limit ice formation to negligible levels (typically <0.5%).

Applications to Tissue-Engineered Products

There are significant challenges for deployment of both preservation methods for tissue-engineered medical products. Vitrification approaches to preservation have some of the limitations associated with conventional freezing approaches. First, both approaches require low-temperature storage and transportation conditions. Neither can be stored above their glass transition temperature for long without significant risk of product damage due to inherent instabilities leading to ice formation and growth. Both approaches employ cryoprotectants with their attendant problems and require competent technical support during rewarming and cryoprotectant elution phases.

The high concentrations of cryoprotectants necessary to facilitate vitrification are potentially toxic because the cells may be exposed to these high concentrations at higher temperatures than in freezing methods of cryopreservation. Cryoprotectants can kill cells by direct chemical toxicity, or indirectly by osmotically induced stresses during suboptimal addition or removal. Upon completion of warming, the cells should not be exposed to temperatures above 0°C for more than a few minutes before the glass-forming cryoprotectants are removed. It is possible to employ vitrified products in highly controlled environments, such as a commercial manufacturing facility or an operating theater, but not in a doctor's outpatient office or in third-world environments.

The cryoprotectants employed for vitrification, in contrast to DMSO or glycerol for freezing, are less well known for preservation applications outside low-temperature biology circles. In particular, formamide, one of the components of the 55% (v/v) vitrification solution consisting of 3.10 M DMSO, 3.10 M formamide and 2.21 M 1,2-propanediol in EuroCollins solution at 4°C (known as VS55), is a known mutagen. Alternatives to formamide with fewer safety risks and potentially easier clinical acceptance are being sought. However, the cytotoxicity of complex cryoprotectant formulations containing formamide is surprisingly much less than the cytotoxicity of single component formulations at the same concentrations.

Storage and shipping temperatures also have a major impact on maintenance of product quality and can result in cell death mediated by ice formation. Degradative processes occur at temperatures warmer than the freezing solution's glass transition temperature (approximately -125°C). Even cells in heart valve leaflets that are frozen slowly can be negatively affected by storage at temperatures warmer than -100°C. It is anticipated that synthetic ice blocker molecules, such as the cyclohexanediols, will be effective in prevention of recrystallization and allow storage of frozen biological materials for longer periods at warmer subzero temperatures. Ice blockers will also allow vitrification at lower, less cytotoxic CPA concentrations. This improved storage capability will facilitate longer shipping times, less expensive shipping methods, and larger cryopreserved specimens.

It is well established that storage and shipping temperatures have a major impact on maintenance of product quality and can result in cell death via ice formation. If storage temperature is sufficiently low (below the glass transition point of the freezing solution [approximately -135°C to -95°C]), little, if any, change occurs in biological materials. Human heart valve leaflets demonstrate retention of protein synthetic capabilities for at least two years of storage below -135°C. Degradative processes may occur at and above the solution's glass transition temperature. For example, it has been shown that cells in cryopreserved human heart valve leaflets are negatively affected by storage at temperatures warmer than -100°C.

One of the major issues for both frozen and vitrified storage of product relates to mechanical forces generated by cooling and warming conditions. Immersion of frozen human valves directly into liquid nitrogen for as little as 5 min may result in tissue fractures. This problem came to light when a hospital-based frozen valve storage system overfilled during an automatic refill cycle. Valves from this accident were discovered to have numerous full-thickness fractures of the valve conduit following normal thawing procedures in the operating room. Adam et al. reproduced this phenomenon experimentally. The rationale for development of fractures appears to relate to abrupt changes in the physical properties of the solidified tissue matrix. Kroener and Luyet described abrupt temperature-dependent changes in aqueous glycerol solutions. Subsequently they reported that the formation and the disappearance of cracks depended on the interaction of several factors, in particular the mechanical properties of the material, the concentration of solute, the temperature gradients, the overall temperature, and the rate of temperature change. Studies of frozen biological materials have also supported the presence of mechanical forces in cryopreserved tissues.

Heat transfer issues are the primary hurdle for scaling up the successes in tissues to larger organs. The limits of heat and mass transfer in bulky systems result in nonuniform cooling and contribute to stresses that may initiate cracking. In fact, the higher cooling rates that facilitate vitrification may lead to higher mechanical stresses. Very little information on the material properties of vitreous aqueous solutions exists. Material properties such as thermal conductivity and fracture strength of vitreous aqueous solutions have many similarities with their inorganic analogues that exist at normal temperatures, e.g., window glass and ceramics. Any material that is unrestrained will undergo a change in size (thermal strain) when subjected to a change in temperature. Calculations of stress in frozen biological tissues have shown that thermal stress can easily reach the yield strength of the frozen tissue resulting in plastic deformations or fractures. We need a much better understanding of mechanical stresses during vitrification and freezing if we are to effectively proceed from long-term biopreservation of simple tissue structures to complex organs in the future.

Issues for the Future

In this overview of biopreservation we have indicated several areas where further research is urgently required, including sterilization methods for allogeneic tissues that permit retention of cell viability, less toxic CPA formulations, better warming methods for large cryopreserved specimens, and the need for a better understanding of the mechanical forces generated by cryopreservation. There are two other topics that we believe should be mentioned in closing that could have a major impact on biopreservation in the future. The first is the development of methods for the intracellular delivery of disaccharide cryoprotectants that are too large to permeate mammalian cell membranes. Success in this area promises new relatively noncytotoxic methods of cell and tissue cryopreservation and leads directly to the second topic of new biopreservation technologies based upon desiccation and freeze drying strategies.

Both conventional freezing and vitrification approaches to preservation have limitations. First, both of these technologies require low-temperature storage and transportation conditions. Neither can be stored above their glass transition for long without significant risk of product damage due to ice formation and growth. Both technologies require competent technical support during the rewarming and CPA elution phase prior to product utilization. This is possible in a high-technology surgical operating theater but not in a doctor's outpatient office or in third-world environments. In contrast, theoretically, a dry product would have none of these issues because it should be stable at room temperature and rehydration should be feasible in a sterile packaging system.

Drying and vitrification have previously been combined for matrix preservation of cardiovascular and skin tissues but not for live cell preservation in tissues or engineered products. However, nature has developed a wide variety of organisms and animals that tolerate dehydration stress by a spectrum of physiological and genetic adaptation mechanisms. Among these adaptive processes, the accumulation of large amounts of disaccharides, especially trehalose and sucrose, are especially noteworthy in almost all anhydrobiotic organisms including plant seeds, bacteria, insects, yeast, brine shrimp, fungi and their spores, cysts of certain crustaceans, and some soil-dwelling animals. The protective effects of trehalose and sucrose may be classified under two general mechanisms: (1) "the water replacement hypothesis" or stabilization of biological membranes and proteins by direct interaction of sugars with polar residues through hydrogen bonding, and (2) stable glass formation (vitrification) by sugars in the dry state. The stabilizing effect of these sugars has also been shown in a number of model systems, including liposomes, membranes, viral particles, and proteins during dry storage at ambient temperatures. On the other hand, the use of these sugars in mammalian cells has been somewhat limited, mainly because mammalian cell membranes are impermeable to disaccharides or larger sugars. Recently, a novel genetically modified pore former has been used to reversibly permeabilize mammalian cells to sugars with significant post-cryopreservation and, to a lesser extent, drying cell survival. Such permeation technologies, which may also include use of pressure or electroporation, may provide some of the most likely opportunities for preservation of tissues in the five- to ten-year vision, either by permitting cryopreservation with nontoxic cryoprotectants or drying. Several methods have been developed for loading of sugars in living cells.

Introduction of trehalose into human pancreatic islet cells during a cell membrane thermotropic lipid-phase transition, prior to freezing in the presence of a mixture of 2 M DMSO and trehalose, has resulted in good cell survival rates. We have found that prolonged cell culture in the presence of trehalose results in significant increases in post-cryopreservation cell survival (patent pending). Human fibroblast transfection with *E. coli* genes expressing trehalose resulted in retention of viability after drying for up to five days. However, it should be noted that most organisms that reach a dried state during dormancy and drought, do so by air drying (not freeze drying), which suggests this may be innocuous to cells under certain conditions. Further, studies of anhydrobiotic organisms may also suggest methods for conditioning mammalian cells for storage by either cryopreservation or drying in the tissue-engineered products of the future.

Concluding Remarks

The emerging fields of tissue engineering and regenerative medicine for living cell-based therapies embody a wide variety of enabling technologies that include the need for effective methods of preservation. Despite significant advances in many of these technologies, it is generally regarded that the basic knowledge and practical know-how needed for the storage of living tissues and complex tissue constructs lags significantly behind. This is reflected by the four key areas of research identified by the US National Institute of Standards and Technology (NIST) in its request for research proposals (1997). These four research areas are automation and scale up, sterilization, product storage, and transportation of product in which substantial technical innovation is required for the development of manufacturing processes.

Concerns for the issues relating to the transition from the laboratory to the market include the major problem of preservation and storage of living biomaterials. Manufacturers and/or distributors recognize the need for maintaining stocks of their products to ensure a steady supply, while the unpredictable clinical demand for specific tissues will necessitate the creation of tissue banks at medical centers. Methods of preservation are crucial for both the source of cells and the final tissue constructs or implantation devices. Tissue preservation technology involves both hypothermic (above freezing) methods for short-term storage, and cryopreservation for long-term banking. Both approaches call for consideration of the cell in relation to its environment and as interventionalists we can control or manipulate that environment to effect an optimized protocol for a given cell or tissue.

In conclusion, tissue preservation as it exists today has been developed empirically and basic research on fundamentals of biopreservation has had restricted impact on the field to this point. In the near future vitrification methods will get a lot more attention, particularly for tissue-engineered products in which immunogenicity is not an issue. In contrast, for allografts, unless the tissue is immuno-privileged, it is unlikely that vitrification will result in significant differences in clinical outcomes. There are, however, two basic research areas that we believe may have a significant impact on tissue preservation in the future. Both are based on lessons we are still learning from nature, namely strategies by which living organisms deal with the environmental temperature extremes to which they are exposed. Through evolution, nature has produced several families of proteins that help animals (e.g., fish and insects) and plants survive cold climates.

These observations led to the hypothesis that naturally occurring antifreeze molecules might be improved upon by synthesis of molecules that will either bind to other ice nuclei domains or upon stable ice crystals. Other organisms naturally accumulate antifreeze compounds such as sucrose and trehalose. Creative methods are required for placement of these compounds within mammalian cells. followed by the development of effective preservation strategies.

5

Molecular Basis of Preservation

As described in the introductory chapter of this text, the field of *biopreservation* is experiencing rapid expansion. This is due in part to the growing interest and success in the areas of personalized medicine and drug discovery. With expanded applications, new demands have been placed on the preservation sciences to provide methodologies that facilitate retention of high degrees of cell viability and function in complex and sensitive systems such as stem cells, cord blood, engineered cells and tissues, etc. In many ways, these demands have pushed the preservation sciences to a limit based on the traditional approaches and parameters guiding the discipline. As a result, the preservation sciences have been forced to morph their focus into the disciplines of cell and molecular biology in search of the next level of answers to drive continued scientific, technological, and methodological development necessary to yield improved outcomes. These efforts have led to a series of recent discoveries related to the molecular response of cells during and following the preservation processes, along with the control of these processes. With these discoveries, we note a paradigm shift leading to new approaches designed to improve preservation. The impact of this paradigm change has had a ripple effect on the discipline of cryobiology as illustrated in the chapter by Gage et al., where these molecular discoveries promise to directly impact one area of cryomedicine.

The underlying phenomenon leading to this paradigm shift has been the discovery of the activation of apoptosis during and following preservation. In 1998 Baust et al. first reported the involvement of apoptosis as one of three distinct mechanisms of cell death contributing to cryopreservation failure: ice rupture, necrosis, and apoptosis. Since that time, a number of studies have been published on cryopreservation-induced apoptosis and molecular-based cell death following cryopreservation. These events identify the putative courses of *cryopreservation-induced delayed-onset cell death* (CIDOD).

This chapter is focused on the discussion of the molecular-based modes, activation sites, and pathways of cryopreservation failure. In addition, the chapter discusses how controlling these factors has and will continue to lead to improved cryopreservation and hypothermic storage outcomes.

Modes of Cell Death

It is now recognized that there are multiple modes of cell death associated with cryopreservation failure and that the cell death processes continue many hours to days following thawing. Detailed descriptions of the preservation process (both hypothermic and cryopreservation) have been provided in the preceding chapters, which mention the importance of understanding a cell's response to the cold in developing the most effective strategy for preservation. The control of a cell's response to low temperature entails a reduction in the stress levels experienced by a cell during the process. In general, increased stress response leads to increased cell death through the activation of apoptotic and necrotic cascades. In cryopreservation control of ice formation and osmotic flux/mechanical response of cells during the freezing process are also important factors. Management of these three modes of cell death (ice rupture, necrosis, and apoptosis) is now believed to play a critical role in preservation efficacy.

Physical Events Related Cell Death

When one thinks of cell death and cryopreservation, failure due to ice-related cell destruction comes to the forefront. The discipline of cryobiology has been dedicated to the understanding, control, and preservation of ice-related cell rupture since the late 1940s when Polge et al. published on the use of glycerol as a cryoprotective agent (CPA) for the successful cryopreservation of avian and bovine spermatozoa. Following this early report, a plethora of studies have been dedicated to preventing intracellular ice formation thereby providing for successful cryopreservation outcome. Numerous mechanistic studies on compounds, which influence ice formation/growth through the "*binding of free water*" (freezable) within a cell, continue to aid our understanding of the

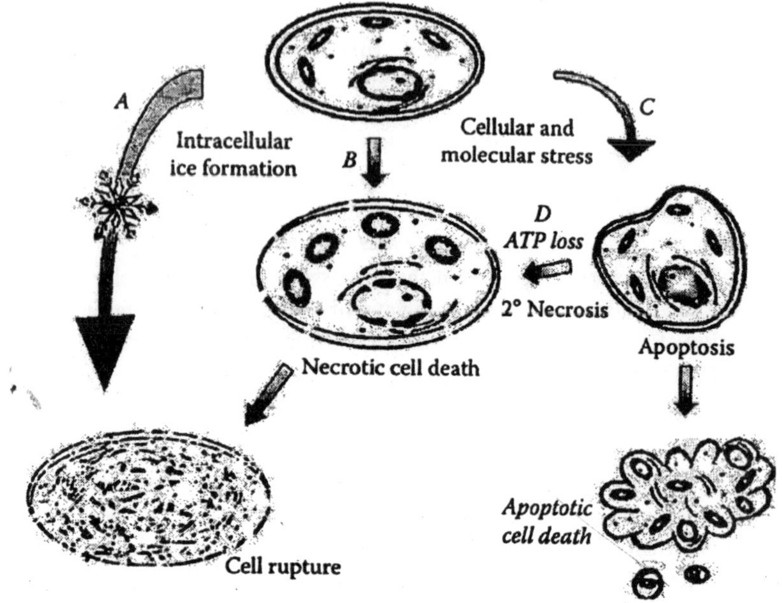

Fig. 5.1. Modes of cell death associated with cryopreservation failure.

cryopreservation process. The process, which depends on the use of agents such as glycerol, dimethyl sulfoxide (DMSO), propanediol, etc., has provided for the effective control and prevention of intracellular ice formation. This focus has played a critical role in advancement of cryopreservation over the past 50 years.

Pathological Cell Death (Necrosis)

While ice-related rupture has received the most research focus, necrotic cell death has been cited in numerous cases of cryopreservation failure. Traditionally, *necrosis* is defined as pathological cell death, a term used essentially to define "*cellular murder.*" Necrotic cell death can be defined as an energy-independent form of cell death characterized by cellular and organelle swelling, loss of membrane integrity, rupture of lysosomes, and random DNA degradation by endonucleases resulting in cell lysis. As a result of the loss of cell membrane integrity, cells lyse, release cytokines, and typically activate immune and inflammatory responses *in vivo*. The initiation and temporal progression of necrotic cell death is often rapid (minutes to hours). Induction of necrosis is typically seen in a response to severe cellular stress and results in a cascade of detrimental intracellular signals resulting in cell death. Necrotic cell death has been shown to be activated by stressors such as ischemia, osmotic shock, sever thermal stress, ionic dis-regulation, toxic agents, etc. Interestingly, many of these necrotic activating stressors are also involved in the complex of stressor responses associated with preservation: *hypothermic* and *cryopreservation*.

Gene Regulated Cell Death

Programmed cell death, or apoptosis, is integral in the homeostatic maintenance of cell number and tissue size in complex organisms. During development, apoptotic cell death plays a role in neuronal innervation of muscles and the deletion of excess cells, such as the webbing between a human's fingers and toes, etc. Apoptotic processes are also a critical line of defense in complex organisms facilitating the daily deletion of damaged cells throughout an organism's lifetime. Kerr et al. first used the term apoptosis (from the Greek meaning "*separation of leaves*") in 1972 to describe cells going through death by what he could only describe as shrinking necrosis. Kerr described apoptosis as a normal physiological process. Following this first report, a distinct field of research devoted to describing, characterizing, and unraveling the associated processes (genes, proteins, cascades, time course, and salient morphological characteristics) has emerged. These studies have led to the characterization of apoptosis as a highly conserved evolutionary process among complex organisms ranging from nematodes to primates. Apoptosis is an energy-dependent form of cell death (requires ATP), which follows a structural route of execution characterized by the activation of specific intracellular proteases, cellular shrinking, maintenance of an intact cell membrane, phospholipid inversion, nonrandom cleavage of DNA into 180 kDa fragments, and formation of membrane blebs (apoptotic bodies). Apoptosis differs significantly from necrosis in that there is no recruitment of a systematic immune response, therefore, no *in vivo* inflammation.

Our knowledge regarding the genetics of apoptosis stems from initial studies conducted on the nematode *C. elegans*. Studies in the area of developmental biology by Horvitz et al. identified the systematic deletion of 131 specific cells during development in *C. elegans* through the activation of specific genes ced-3 and ced-4. Further, Hengartner et al. showed that through the action of the gene ced-9, programmed cell death could be prevented. Following the identification of these genes and their processes in the nematode, multiple homologs to the ced-3, 4, and 9 genes were found in higher eukaryotes including mice, rates, and humans. The cell death gene ced-3 was found to be homologous to interleukin IB converting enzyme (ICE). ICE, a cysteine protease, is common to many organisms and is therefore indicative of a conserved programmed cell death mechanism throughout the animal kingdom. Further, investigations have shown that in mammalian systems, a series of ICE like cysteine proteases, termed caspases (cysteinyl aspartate-specific proteases), plays a role in apoptotic cell death (ICE = caspase 3).

Apoptotic cell death is characterized by three stages: initiation, execution, and termination. During each stage, a series of specific events is activated as part of a complex and elegantly choreographed cascade leading to cell death. Progression of the cascade through each stage, as well as through the substages, requires energy input (ATP) at checkpoints throughout the process. During the progression of apoptosis, if energy (ATP) is unavailable, cells shunt to a necrotic cell death pathway. This shunting has been termed *secondary necrosis* and is discussed in the transitional cell death section of this chapter. Apoptosis has been shown to be initiated by a series of stresses including, but not limited to, radiation, cytotoxic agents, nutrient deprivation, anoxia, growth factor withdrawal, temperature, excess or diminished gene products (*apoptotic* and *homeostatic*), etc. Following induction at any number of organelles, apoptosis proceeds through a cascade of events including, but not limited to, caspase activation, mitochondrial release of cytochrome C, cell cycle arrest, externalization of membranous phosphatidyl serine, alterations in gene expression, etc. This sequence of events leads the cell through the execution stage into the termination stage where DNA is cleaved by exonucleases into ordered fragments, membrane blebs and apoptotic bodies form and the complete systematic disassembly of the cell is accomplished.

Time to Die

As discussed, apoptosis is an intricate part of the everyday life of an organism. During development, the apoptotic process helps to "*mold*" organisms in many differing ways—from the simple tissue level, to organs, to the entire appearance of the organism. As previously mentioned, in *C. elegans* cells die at programmed intervals in order to "mold" the organism. At the mammalian level, apoptosis too plays an equivalent role in the maintenance and control of both life and death of the organism through the targeted removal of "*unnecessary*" cells. For example, it is well documented that during natal human development, the fetus possesses webbing between its digits. However, upon birth, in most cases, the webbing is absent. The selective removal of this webbing prior to birth can be attributed to selective, developmental programmed cell death. The same can be said for the observed removal of fetal tail buds. Developmental apoptosis does not simply involve the elimination of vestigial features of a fetus, but also is an intricate process in the development of the human nervous system, playing a role in brain cortical development and neuronal innervation of tissues and organs. Following fetal developmental cessation and birth, apoptosis continues to play a role in human survival. The apoptotic processes are responsible for the continual maintenance of the cellular components of all tissues and organ systems within the body. Within organ systems such as the skin or liver, apoptosis plays a daily role in the turnover and regeneration processes necessary to maintain viable and functional organ systems. With apoptosis playing such a predominant yet selective role in organismal homeostasis, the questions of "when cells know it is time to die and how these processes are initiated within the cell" have become central to shaping today's research in this area.

A definitive answer to the question of "when to die" is proving difficult to obtain. The induction of the apoptotic processes by selective cells within organ systems can be mediated by a number of differing stimuli ranging from cellular "old age" to lethal genomic DNA errors to cellular stresses (e.g., hypothermia, hypoxia hypersensitivity, nutrient deprivation, exposure to toxins, etc.). These death-initiating stimuli experienced by a single cell or group of cells within an organism may result in the induction of the apoptotic process through a number of differing pathways. How exactly an individual cell processes the information on "when to die" is actively being researched. With new information, one common theme seems to resonate throughout the field of apoptotic research: cells do die and we are still not sure why.

Induction of Apoptosis

The induction of apoptosis can result from a number of different stress factors experienced by a cell. Initiation of apoptotic cascades occurs at one of three primary locations within a cell: the cell membrane, the mitochondria, or the nucleus. Each of the initiation points results in the activation and progression of cell death through a distinct pathway. Despite activation of a distinct pathway by the initiating organelle, once activated there is a series of feedback, amplification, and cross-talk biochemical signaling loops, which are subsequently activated and result in the potentiation of the apoptotic signal. This cross-talk and amplification is believed to have evolved as a redundant biochemical system to help assure ordered cell death following the initiation of apoptosis. As will be discussed in the Targeted Apoptotic Control (TAC) section, it is this feature that makes apoptotic control in cryopreservation difficult.

There is a wealth of knowledge surrounding the signal transduction cascades associated with the progression of apoptosis from all three initiating organelles. Researchers have shown that there is a distinct set of signaling proteins associated with apoptotic initiation from each organelle, all of which typically feed into the proteolytic caspase cascade leading to cell death.

Membrane Receptor Mediated Apoptosis

One common location for the initiation of apoptosis is that of the cell membrane. Every cell has a series of protein receptors on its surface designed to initiate the apoptotic cascade. These receptors include several members of the TNF superfamily including FAS, TNF1 and 2, TRAIL, TRANCE, etc. These surface receptors have been shown to be pleiotropic mediators of an immune response as well as apoptotic program initiators through ligand binding.

Activation of one of the TNF superfamily surface-receptor proteins leads to the activation of a series of downstream signal transduction (protein) messages. One such receptor, TNF1, when activated, results in the recruitment of the intracellular protein TRADD. TRADD binds to the activated TNF1 complex on the intracellular side of the plasma membrane. Once bound and activated, TRADD acts as a signal and recruitment adaptor molecule for the protein FADD (Fas activated death domain), which in turn acts as a recruitment signal and adaptor protein for procaspase 8 (inactive caspase 8).

Following binding to FADD, procaspase 8 undergoes proteolytic processing, resulting in the activation and release of active caspase. Once activated, caspase 8, through a specific pathway, activates caspase 3, 6, and 7. Caspase 7 in turn acts as a signal amplification molecule to recruit and activate additional pathways, and results in cleavage of cellular DNA followed by cell death.

Mitochondrial Induced Apoptosis

As much as any other organelle, the mitochondrion plays a critical role in both cell survival and death. The mitochondria within a cell are responsible for generation of most of a cell's energy through oxidative respiration. While critical to cell survival, the mitochondrion also serve as a primary activation site for apoptotic cell death. Accordingly, there are several genes and proteins associated with the mitochondria that regulate its functional status of a life sustainer or terminator. The mitochondria are now recognized as key to both necrosis and apoptosis. The decision by a cell to launch apoptosis can be, in part, dictated by the cytoplasmic ratio of the antiapoptotic to pro-apoptotic proteins. Most of the known anti-apoptotic and pro-apoptotic proteins are directly related to the mitochondrial Bcl-2 family. Bcl-2 was originally found at the chromosomal translocation in follicular lymphomas and subsequently determined to inhibit apoptosis. Experiments in the nematode *C. elegans* determined that the apoptosis-inhibiting gene, ced-9, was homologous to Bcl-2 and served a similar function, suggesting that apoptosis prevention may be of universal importance.

The family of proteins most commonly associated with the mitochondria is the Bcl-2 family, which consists of both pro- and anti-apoptotic members. The mechanisms by which these pro- and anti-apoptotic proteins work are still not fully understood. The anti-apoptotic Bcl-2 family members may work through dimerization with other Bcl-2 family members, inhibition of caspase cleavage (activation), or phosphorylation of selected proteins. Bcl-2 may also work by blocking the mitochondrial permeability transition pore that is thought to initiate apoptosis triggered by the release of cytochrome C through voltage-dependent anion channel (VDAC). Cytochrome C, in turn, may activate procaspases, leading to the demise of the cell. The proteins comprising the VDAC are thought to work together to permit the increase in permeability of the mitochondrial inner membrane to solutes with a molecular mass below 1500 Da that include superoxide anions that leak to the cytosol. An inward flux of these molecules through the VDAC results in the loss of proton motive force in the mictohondria, triggering apoptosis and/or necrosis.

Bcl-2 and Bcl-X_L family members are anti-apoptotic while Bax, Bad, and Bid are pro-apoptotic. These proteins act by regulating (or deregulating in the case of pro-apoptosis) the mitochondrial permeability transition pore (mPTP), mitochondrial transmembrane potential, and the subsequent release of cytochrome C from the inner mitochondrial space. When opening the mPTP the result is a decrease/loss in the mitochondrial transmembrane potential and release of cytochrome C into the cytosol where it binds (heterodimerizes) with the molecule APAF-1. This heterodimerization results in the recruitment of procaspase 9, forming a complex known as the apoptosome, resulting in the activation and release of caspase 9. Following activation, caspase 9 (an upstream caspase) acts as an activator of downstream execution caspases (caspases 3, 6, 7), as well as a feedback-signaling molecule causing the release of additional cytochrome C and transcriptional up-regulation of procaspase 9 mRNA. As described in membrane-mediated apoptosis, activation of caspase 3, 6, and 7 results in a series of degradation events for target molecules such as PARP (poly-ADP ribose-polymerase), DNA, ICAD, etc., leading to cell death.

Nuclear-induced Apoptosis

Another primary site of apoptotic induction is that of a cell's nucleus. Cell stressors that act on nuclear targets often result in the initiation of an apoptotic response. Stress factors such as UV radiation (direct DNA damage), growth factor deprivation, deregulation of cell cycle processes (DNA synthesis, alterations in cell cycle regulation, and at protein levels, disruption of cytoskeletal protein arrangement, etc.), oxygen free radicals, toxins, steroids, etc. have all been shown to initiate a nuclear-mediated apoptotic response. Most, if not all, of the nuclear-mediated apoptotic response stressors result in some form of DNA damage (breaks, nucleotide misincorporation, genetic recombination). Alterations in DNA integrity typically result in the activation of p53 (a DNA proofreading protein), which in turn signals cell cycle arrest and initiates a series of signaling cascades leading to either DNA repair or apoptosis. The decision between activation of repair processes vs. cell death progression is complex and involves a host of signal-mediating proteins related to the intensity of the stress, extent of DNA damage, and the availability of ATP to effect repair. In extreme cases of DNA damage, the signaling pathway initiated by p53 has also been shown to shunt the cell from apoptotic to necrotic cell death when energy levels are low.

Transitional Cell Death

Historically, molecular-based cell death has been thought to proceed through either an apoptotic or necrotic path. Apoptosis has been viewed as "more of a molecular response" with necrosis considered "less molecular" at the intracellular signaling level. However, our view of the molecular cell death landscape has evolved substantially over the past few years to suggest that apoptosis and necrosis represent more of extremes on a continuum of molecular-based cell death. Apoptosis is now viewed as a mode of cell death that can present itself in many forms. For instance, Bras et al. have suggested that three types of apoptosis occur. Type I is the conventional view of apoptosis that does not involve lysosomes but does rely on caspase activation. Presently, over a dozen caspases are recognized, many of which are linked together. Type II, by contrast, is characterized by lysosomal-linked autophagocytosis. Type III is lysosomal-independent, necrosis-like apoptosis marked by swelling of intracellular organelles. These three types may be linked by the mitochondria and the critical role this organelle may play in switching between all three death modes. Many of the caspases are linked to the mitochondria and appear to play roles both in apoptosis and necrosis. Essentially, it is now believed that in many cases when a cell commits to death, an apoptotic response is activated and proceeds to the point of cellular execution (classical apoptosis) or to a point where the stress becomes too great or energy reserves (ATP levels) too low for continuation of apoptosis. At this point the cell has committed to death and shunts from apoptosis to necrosis for the completion of the cell death process (secondary necrosis). The ability of cells to switch from apoptotic to necrotic cell death and even back again was first demonstrated in 1997 by Liest et al. In this study, Jurkat cells were enticed into apoptosis, and then depleted of energy, which led to a shift in cell death characteristics from apoptotic to necrotic. Interestingly, when energy substrates were added back to the system, cell death characteristics reverted back to apoptotic. However, there was a point at which this reversal was no longer possible. This transitional cell death has been demonstrated in a number of subsequent studies and has provided a basis for the cell death continuum concept emphasized in this chapter. Common stressors such as nutrient deprivation, DNA damage, cytokine exposure, cytotoxic agents, oxygen deprivation, ionic imbalance, etc. have all been shown to result in the activation of both apoptosis and also necrosis in a multiplicity of cell systems based on the relative degree of the stress experienced by the cell. Another situation where observations of transitional cell death have been noted is that of genomic and proteomic alterations in cell death pathways. There are many examples of gene mutations—alterations in expression or deletion—resulting in the inability of a cell to progress properly through classical apoptotic cascades, thereby switching to necrosis. One such example of "*atypical*" apoptosis is observed in the human prostate cancer cell line PC-3. PC-3 cells are deficient in the DNA proofreading, apoptotic initiator protein p53. This mutation results in a reduction in the ability of this cell type to undergo apoptosis. Despite this mutation, PC-3 cells can progress through both apoptotic and necrotic pathways, which would typically involve p53 by recruiting additional apoptotic and necrotic response elements to continue cell death. Other examples are seen when there is an inhibition of caspase activity in cells undergoing apoptosis. In many cases these cells simply shift from apoptotic to necrotic cell death when the necessary caspase becomes unavailable.

The transitional nature of cell death pathways in response to similar stressors makes for an extremely complex environment to characterize when it comes to situations where multiple stressors are involved, such as cryopreservation and hypothermic preservation.

Cryopreservation-Induced, Molecular-Based Cell Death

It has only recently been discovered that cell death following cryopreservation is linked with apoptotic and secondary necrotic mechanisms. Upon reflection and consideration of the stresses associated with cryopreservation, it now seems intuitive that apoptotic processes should be involved with cryopreservation failure. In 1995, Jurisicova, et al. reported on observations of apoptosis contributing to cellular demise following cryopreservation in pre-implanted human embryos. The authors identified programmed cell death (PCD) as a contributor to post-cryopreservation embryo demise and hypothesized that "the PCD was due to either natural incidence of lethal chromosomes in the human population or due to the *in vitro* fertilization and culture conditions." Later, Borderie et al. reported on the observation of a discrepancy (20–54%) in cell survival following frozen storage (-80°C) of human keratinocytes.

The discrepancy was noted when results were obtained through assessment of viability using trypan blue and flow cytometry. The authors also noted the presence of apoptotic cells within the culture following a 24-hour recovery period. Additionally, the presence of apoptotic cells following transplantation (2–14 days) of cryopreserved allograft heart valves has also been reported. Despite these observations, it was not until 1998 that a first report directly linking apoptosis to cell death following cryopreservation was published. The discovery of apoptotic involvement was found through electrophoretic analysis of DNA from cell populations one to two days post-thaw. These DNA gels demonstrated that apoptosis was occurring as a result of cryopreservation through the appearance of the hallmark 180–200bp DNA fragment bands.

Interestingly, it was seen that the hallmark DNA characteristics of apoptotic (DNA banding or ladder) and necrotic (DNA smear) cell death were not detectable until many hours to days post-thaw, revealing a significant temporal component to cell death following cryopreservation. Since this report, there have been numerous studies looking to identify apoptotic involvement in cryopreservation failure.

With this increase in activity, there have been numerous reports showing apoptotic involvement as well as a handful claiming the contrary (not finding apoptosis) in their respective systems.

Apoptosis in Cryopreservation

Apoptosis has now been documented in a wide variety of cellular systems following cryopreservation. Since 1998, studies identifying post-thaw apoptosis have begun to appear from numerous systems including renal cells, fibroblasts, hepatocytes, peripheral blood mononuclear cells, cord blood, spermatozoa, oocytes, ovarian tissue, vascular tissue, etc.

In 2000, Fowke et al. and Baust et al. separately reported on apoptosis following cryopreservation in PBMC and renal cells, respectively. In the following year Fu et al. and Yagi et al. reported on the involvement of apoptosis following cryopreservation in mouse and porcine hepatocytes, respectively. Additionally, Schuurhuis et al. and Lund et al. documented apoptosis in PBMCs following thawing. Further reports in the gamete cryopreservation area have documented the presence of apoptotic cells following thawing in sperm, oocytes, and ovarian tissue. The presence and contribution of apoptosis has also now been reported in red blood cells, cornea, stem cells, cord blood, lymphocytes, and oocytes. These reports as well as others continue to solidify the foundation of molecular-based cell death following cryopreservation as a universal phenomenon influencing cryopreservation outcome.

Cryopreservation-induced, Delayed-onset Cell Death

From a review of the literature on the type, time course, and extent of cell death associated with cryopreservation, one can now conclude that molecular-based cell death (apoptosis) plays a critical role in the outcome of many, if not all, cellular systems. One critical point underlying this statement (and interpretation of the literature base) is that of the temporal component of post-cryopreservation cell death and the assay system used for evaluation. Simply stated, evaluation of cell populations within a few hours post-thaw does not allow for the identification of the full extent of apoptosis or necrosis due to the time course of the cell death machinery regardless of the assay system utilized. The manifestation of molecular-based cell death may take many hours to days to occur following thawing. It is this temporal component that continues to elude many investigators seeking to identify and characterize molecular cell death following preservation. In 2001, a report detailed the timing of cell death following cryopreservation,

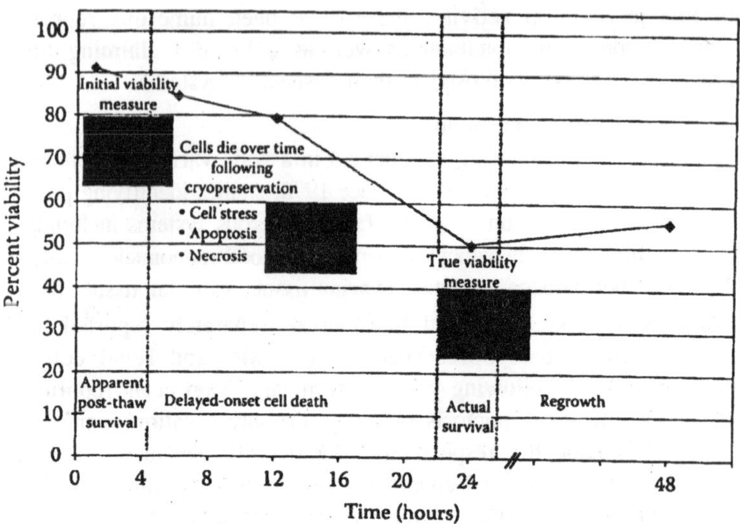

Fig. 5.2. Effect of cryopreservation-induced, delayed-onset cell death on cell survival following thawing

terming the phenomena *Cryopreservation-Induced, Delayed-Onset Cell Death* (CIDOCD). This study showed that following thawing there was a delay in the peak of necrotic activity until 6 hours post-thaw with a subsequent peak in apoptosis at 12 hours post-thaw.

The measurements of apoptotic and necrotic activity were based on early stage event indicators, thereby identifying cells that were indistinguishable from living cells based upon typical viability indicators (i.e., trypan blue). With the progression of the cell death cascades, it was not until 24 hours post-thaw that a nadir in cell survival was observed. From this study, it was documented that the timing of analysis for assessment of cryopreservation failure is critical. With the identification of the temporal component of cryopreservation failure, a series of investigations were launched to uncover the path of molecular cell death progression. In 2002, Baust et al. reported on a genomic response within human fibroblasts following cryopreservation contributing to this temporal component. This study reported that following thawing there was an increase in transcriptional activity of key apoptotic enzymes (caspase 3, 8, 9) in a delayed manner with a peak in transcriptional activity of caspase 8 and 9 at 12 hours post-thaw and caspase 3 at 18 hours post-thaw.

Further analysis of caspase 3-proteolytic activity revealed an extended plateau in activity 21 to 36 hours post-thaw. This gene-mediated response resulted in an observed nadir in cell survival at 24 hours post-thaw with continued cell death out to 48 hours post-thaw. Concurrent with this study in fibroblasts, Vogel et al. reported on the proteolytic activation of caspase 3 following cryopreservation in a renal model. This study demonstrated substantial alterations in the cellular-outcome proteolytic activity immediately post-thaw (6 hours) followed by a subsequent increase in activity over 12 to 18 hours post-thaw with a return to a basal level by 24 to 36 hours. Schmidt-Mende et al. independently substantiated the observation of post-thaw protease activation in a bone marrow cells model. They reported on a high level of intrinsic proteolytic activity following preservation leading to the cleavage of various apoptotic-signaling proteins.

These studies further demonstrated that protease activity was blocked by the inclusion of a protease inhibitor in the freeze medium. These studies demonstrated that improved cryopreservation outcome was possible through the modulation of the apoptotic process via caspase inhibition during and following the cryopreservation interval. To further implicate caspase activation following cryopreservation, Paasch et al. reported on the activation of caspase in human spermatozoa and Yagi et al. did so following thawing in a porcine hepatocytes model. Based upon the reports of specific protein upregulation, Vogel, et al., more recently reported on an increase in proteolytic activity of numerous proteins following cryopreservation in a human fibroblast model.

Effects of Cryopreservation on Cell Function

While a substantial literature base exists detailing the effects of post-cryopreservation apoptosis, there are, nonetheless, recent reports refuting these claims. In 2002, Ricco et al. reported that cryopreservation of PBMC did not result in increased levels of apoptosis 24 hours post-thaw. These findings are in contrast to the previously cited studies. Further, in a recent study by Baust et al., it was reported that both apoptosis and necrosis play a significant role in cell death following cryopreservation. Comparison of the two studies revealed that the temporal component of PBMC death played a critical role in guiding the interpretation of results (no finding of apoptosis). In the Ricco et al. study, PBMC analysis was conducted immediately post-thaw and 24 hours later. In the Baust et al. study, PBMC necrosis and apoptosis was found to peak at 4 to 8 hours, respectively, with minimal levels of either form of cell death noted at 24 hours. Comparison of these two studies serves as an example of the influence that the temporal components and cell type specificity can have on interpretation of results. Subsequent studies have shown that the delayed molecular effects following thawing extend beyond that of cell death but impact function as well. Overall function of cellular systems following cryopreservation had been an issue often neglected in the past. The literature contains numerous reports citing high retention of cell viability and function post-thaw. However, in-depth examination of the literature reveals that there is a substantial reduction in function of many cells post-thaw. This effect has been documented in hepatocytes, pancreatic islets, cardiac cells, blood cells, and stem cells, to name but a few. In 2000, Fowke et al. described the use of early apoptotic events as a means of assessing sample quality in peripheral blood mononuclear cells (PBMC). Abrahamsen et al. applied a similar approach by using flow cytometry to assess apoptosis and necrosis levels in PBMC samples following cryopreservation as a means of determining cell concentration and dosing parameters in cancer patients. These authors also reported that cryopreservation affected the level of $CD34^+$ expressing cells in PBMC samples. Later de Boer et al. showed that cryopreservation resulted in an impairment of function in $CD34^+$ cells reducing their effectiveness in stem cell grafts. Reports in the area of gamete cryopreservation have also described decreased function. Studies in cryopreservation of sperm now link cryopreservation and molecular responses along with a loss of acrosomal and motility functions. More recently, studies have begun to detail the effects of cryopreservation on the biochemical functionality of cell systems such as hepatocytes and cardiomyocytes as well as the improvement in cell function through newly developed approaches to cryopreservation. Baust et al. in an analysis of the impact of cryopreservation on the cellular proteome found that there is an effect on a cell's phenotypic fingerprint many hours to days post-thaw. Van Buskirk et al. also recently discussed this effect on the cellular proteome. These examples provide a mere sampling of the downstream effects cryopreservation may have on cellular function.

Initiation of Cryopreservation-induced Molecular Death

Much attention has been paid to identifying and quantifying molecular cell death following cryopreservation, but few detailed investigations into the initiating stresses have been conducted. Inherent in the process of cryopreservation is the exposure of cells to stressors, many of which can initiate a molecular death response in conjunction with one another. Many of these factors were recently reviewed and include metabolic uncoupling, production of free radicals, alterations in cell membrane structure and fluidity, deregulation of cellular ionic balances, release of calcium from intracellular stores, osmotic fluxes, and cryoprotective agent exposure. This list of stresses associated with cryopreservation is by no means complete but serves as a guide to the complexity of stressors, stress response mechanisms, and influence on cellular initiation sites. In many cases, the accumulation of numerous sublethal stressors during the preservation process results in activation of apoptosis followed by a shift to secondary necrosis due to a lack of energy and continual "build up" of stressors within the cell. In an attempt to gain insight into the effect of the various stressors associated with cryopreservation, recent studies have focused on the various initiation sites of apoptosis. These studies shed light onto the pathways of molecular cell death associated with cryopreservation.

Membrane Events

The effect of the freezing process on the cell membrane has been studied extensively from the ultrastructural prospective. Studies detailing the increase in viscosity of the cell membrane in response to low temperature are common. Further studies have shown that the membrane becomes more permeable at low temperatures due to channel formation in the lipid bilayer. Steponkus et al. reported on the aggregation of proteins and the formation of circular lipid rings within the lipid bilayer of the cell resulting in irreversible damage and subsequent cell death.

Recently, studies have turned toward the analysis of molecular-induced events associated with the cell membrane following cryopreservation. As a result, there is now a growing yet limited base of knowledge in membrane-mediated responses. We first implicated the involvement of the cell membrane in cryopreservation-induced apoptosis in a 2002 study, which found that during post- thaw recovery there was a transcriptional upregulation of caspase 8. Further studies have shown that following thawing there is an alteration in the protein levels of Fas and FADD along with the activation of caspase 8. Additionally, studies from the field of cryomedicine have shown the activation of TNF and TRAIL pathways in response to freezing. Most recently, our group has noted that the inhibition of FAS activation during cryopreservation results in an increase in cell survival. In contrast, Bischof's group has shown that through the activation of the TNF receptor, the efficacy of cryodestruction can be increased. These studies, taken together, argue for a strong involvement of the activation of membrane-induced molecular cell death signaling following low-temperature exposure.

Mitochondrial Response to Low Temperature

As described earlier, the mitochondrion plays a pivotal role in the control and initiation of apoptosis. In the area of low-temperature stress/response, several studies have now linked the mitochondria to the initiation of apoptosis following preservation. In 2001 our group reported on the transcriptional activation of caspase 9 following cryopreservation. Ahn et al. extended these studies by reporting on mitochondrial dysfunction in two-cell mouse embryos following cryopreservation. Subsequent to this report, a series of studies appeared investigating mitochondrial involvement in both cryopreservation and hypothermic storage cell death. Vogel et al. showed that there was an alteration in the Bcl-2 and Bax levels following cryopreservation resulting in a shift toward intracellular pro-death cell death signaling in a human fibroblast model. Clarke et al. reported a similar finding in prostate cancer cells in response to freezing. Lastly, Anderson et al. published on the translocation of Bax from the cytosol to the mitochondria following low-temperature exposure in a hypothermia model. Investigations into mitochondrial involvement have now shown that following low-temperature storage there is a loss of mitochondrial transmembrane potential as well as a release of cytochrome C, the activation of caspase 9, and subsequent apoptosis. A number of additional studies exist on mitochondrial involvement (especially in the area of hypothermic storage), but from the above studies it is clear that the mitochondria play a crucial role in activation of molecular-based cell death following preservation.

Nuclear Initiation of Apoptosis

While a literature base exists for the mitochondrial and membrane-induced apoptosis associated with cryopreservation, there is a much smaller body of information detailing nuclear involvement. To date, most investigations have focused on examination of nuclear events as an identification of molecular cell death utilizing assay systems such as TUNNEL, DNA gel electrophoresis, comet assay, and western blotting of nuclear proteins. These studies have shown a disruption in DNA integrity, cleavage of nuclear proteins such as PARP, increase in xIAP, CAD activation, etc. All of these events have been studied in the context of downstream effects of the apoptotic process activated following cryopreservation. Specific studies on actual nuclear events that lead to the activation of apoptosis are limited. With that said, there are studies demonstrating DNA damage in spermatozoa following thawing and subsequent cell death. Further, a recent study by Clarke et al. utilizing cDNA microarray analysis in a cryotherapy model showed a significant genetic response to freezing within a few hours post-thaw. While this study did not focus specifically on the nucleus as an initiation site, it nonetheless demonstrated a large nuclear-based response to freezing leading to a number of cascade activations including those of nuclear-initiated apoptosis. With the explosion of microarray analysis and the plethora of gene activation data, it is only a matter of time before a more in-depth understanding of nuclear-initiated apoptosis in cryopreservation failure will be described.

Control of Cryopreservation-induced Molecular Response

With the discovery of molecular responses in cells to the preservation process, there have been numerous attempts to control these events in an effort to improve preservation outcome. These approaches vary from alteration in solution (cryoprotectant carrier media) design to addition of protective agents for targeted control of apoptosis. Of these attempts the alteration in preservation solution design has proven most beneficial.

Next-generation Cryopreservation Solutions

The most significant shift in the approach to improve cryopreservation methodologies in recent years is that of carrier media design. Traditional cryopreservation media formulations consist of a basal culture media with serum protein and DMSO supplementation. While providing for physical protection through the DMSO and proteins components, the basal solutions do not provide for control/modulation of the molecular response of cells to preservation stresses. These solutions fall short by failing to control alterations in solution pH, cellular generation of free radicals, energy deprivation, etc. Further, culture media based solutions fail to provide the appropriate ionic environment necessary for cell survival and reduction of cellular stress during preservation. These media are historically, exclusively "*extracellular-like*" with regard to ionic concentrations (high Na^+, low K^+). These properties of the preservation solution do not adequately protect at a cellular and molecular level, thereby exacerbating cell death.

In contrast to traditional carrier media, a new generation of cryopreservation solutions is designed around the concept of an "intracellular-type carrier media" (low Na^+, high K^+). Additionally, these solutions contain buffers, which have an extended dynamic thermal

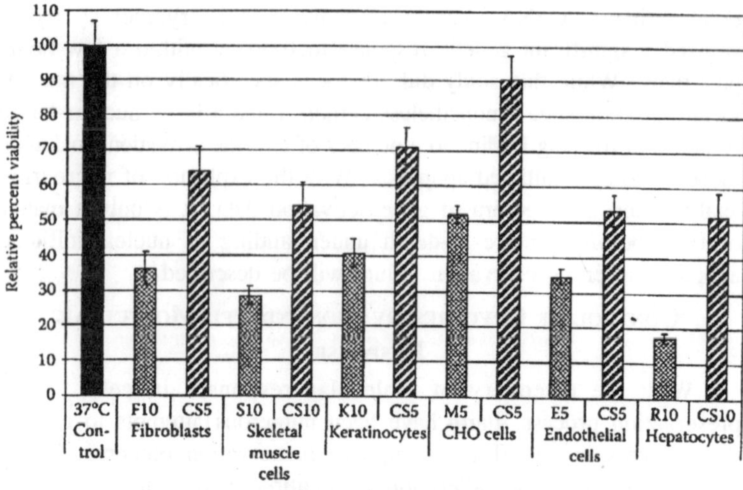

Fig. 5.3. Comparison of the effect of extracellular vs. intracellular-like cryopreservation solutions on 24- hour post-thaw viability in model cell systems.

range of activity, supplemental free-radical scavengers, energy substrates, etc. In fact, the formulation of these cryopreservation solutions has taken the lead from the field of hypothermic storage solutions to provide for more effective cryopreservation. Successes with this approach have been detailed in a series of reports in cellular systems such as hepatocytes, cord blood, stem cells, PBMC's, fibroblasts, keratinocytes, blood vessels, engineered tissue, etc. In most of these studies evaluation of the cryopreservation media CryoStor was conducted, which yielded significant improvement in cell survival, function, and growth. In many cases a doubling of cell number was noted at the nadir of cell survival when contrasted with DMSO plus cell culture media. Interestingly, in most of these studies, the improvement in viability and sample quality was not noted immediately post-thaw. It was not until the molecular-based events fully manifested that the significant improvement was observed. Studies have shown that the basis of the observed improvement in cell survival and function was due to a direct reduction in the level of both apoptosis and necrosis during post-thaw recovery. This relationship has been demonstrated in a number of studies, which have detailed a direct reduction in apoptotic and necrotic cell number, transcriptional up-regulation of caspases, shift in Bcl-2/Bax ratio, and the extent of DNA damage. These studies have all shown that the use of new-generation cryopreservation solutions is significantly changing cryopreservation outcomes.

Targeted Apoptotic Control

While next-generation solutions have provided for marked overall improvement in cryopreservation outcome, another area of interest is that of Targeted Apoptotic Control (TAC). The identification of the involvement of apoptosis in cryopreservation failure has led to numerous studies investigating the feasibility and effectiveness of TAC. In 2000 our group first reported on the attempt to control the activation of apoptosis following cryopreservation through caspase inhibition in a renal cell model. This study reported that TAC, in combination with CryoStor solutions, could markedly improve cell survival. A subsequent study demonstrated that through TAC both the levels of post-thaw apoptosis and necrosis could be significantly reduced. These findings provided a basis for a hypothesis detailing transitional cell death playing a role in cryopreservation failure. Yagi et al. have reported on the benefit of TAC in improving hepatocyte cryopreservation, a significant investigative milestone.

Following this study, a number of additional reports have been published on the incorporation of caspase inhibitors into cryopreservation media to effect improved cell survival via TAC. Recently, Robilotto et al. described the benefit of calpain inhibitors during cryopreservation to improve cell survival. While a number of studies have been reported, to date none have proven practical for broad-based application in the cryopreservation sciences.

Enabling Tissue Cryopreservation

Generally speaking, the emphasis in the cryopreservation sciences has been on cellular systems both at the physical as well as molecular levels. Translation of these approaches into the area of tissue cryopreservation has been formidable. This area includes both tissue freezing and vitrification. Here, we choose to focus on tissue freezing and studies involving molecular findings related to cell death and function. The challenges faced in the cryopreservation of cells translate directly into tissue systems compounded by the existence of cell-to-cell interactions, diffusion and permeability limits, nonuniform heat transfer (both during cooling and warming), etc. These factors may add to the stress levels experienced by a cell and result in even greater cell death following cryopreservation.

At the molecular level it is believed that sheer stress and strain experienced by cells due to volumetric changes and ice formation, coupled with a cell's anchoring to the extracellular matrix, result in the activation of a stress response pathway leading to apoptosis.

Additionally, it has been hypothesized that due to cell-to-cell connections in tissues, the transmission of cell stress and death signals can "flow" more freely between cells, thereby leading to increased cell death through signal amplification. It is believed that these physical and molecular signals/events work directly in concert with one another, making tissue cryopreservation difficult.

Currently, there are a number of tissucs being cryopreserved to meet clinical and research needs. Systems routinely cryopreserved include liver slices, heart valves, blood vessels, cartilage, skin, and numerous engineered tissues. Despite the "*routine*" freezing of these systems, tissue quality is quite poor. The process of tissue cryopreservation differs little from that of cell cryopreservation except for the exter ion of the initial 4°C incubation interval. These approaches typically result in "*viable tissue*" (maintenance of tissue integrity) used in applications such as heart valve transplant, vascular replacement therapy, etc., in which the tissue primarily provides a "living" scaffold that cells from other parts of the body can infiltrate over time.

A number of studies have also reported on molecular events associated with cryopreservation of tissue. Liu et al. reported that cryopreservation had only minor effects on mouse ovarian tissue including *in vitro* follicular maturation and gene expression. This study documented that those follicles survived the preservation process and were able to mature in a normal fashion. They also showed little effect on gene expression after recovery. Matsushita et al. reported on the translation of molecular studies of hepatocyte cryopreservation on the development of bio-artificial liver assist devices and reported on the role of apoptosis and the cryopreservation process and overall device function. Villalba et al. and Pascual et al. described the effects of cryopreservation in vascular tissue models (heart valves and arteries, respectively), implicating delayed cell death (*apoptosis* and *necrosis*) as playing a role in tissue destruction following cryopreservation.

More recently, a series of studies that revolve around the development of new, molecular-driven approaches to the cryopreservation of tissues has emerged.

Snyder et al. recently reported on the application of molecular approaches in the cryopreservation of rat aorta. This study showed that cryopreservation of aortic segments utilizing traditional approaches resulted in a structurally intact nonviable tissue. However, when utilizing an intracellular-like preservation solution, tissue cryopreservation outcome was markedly improved. Specifically, aortic segments preserved in CryoStor CS 5 or CS 10 resulted in a marked increase in viable cell number along with additional cell growth. The appearance of organized cellular outgrowth was only observed in the new-generation solutions.

The ability to further increase tissue cryopreservation efficacy using new approaches has also been detailed in a tissue-engineered skin equivalent model. This study reported on the benefits of altered solution design in combination with TAC focusing on either membrane-mediated and mitochondrial-mediated apoptotic events. The data revealed a marked increase in tissue viability and integrity compared to traditional media + serum + DMSO approaches.

While the area of molecular-based tissue cryopreservation remains in the early stages, the studies to date have provided a solid platform for continued experimentation. Coupled with the abundance of data available on the hypothermic storage of tissues and organs, it is clear that advances in the cryopreservation of tissues and possibly organs will continue to progress in the coming years.

Molecular Response to Hypothermic Storage

This chapter has focused primarily on the discipline of cryopreservation with reference to hypothermic storage throughout. Previous study provides an in-depth discussion on the fundamentals of hypothermic storage, with an emphasis on preservation solution design. Many of the topics discussed within that chapter have had direct implications on cryopreservation. As such, many of the principles of molecular-based cell death discussed within the context of cryopreservation also apply directly into hypothermic storage. In fact, studies into molecular-based apoptotic and necrotic cell death in hypothermia are more advanced than in cryopreservation. Studies by Mathew et al., Rauen et al., Ohno et al., Rugo et al., Huet et al., and Hansen et al., to name a few, have detailed the activation and effects of stress-induced apoptosis and necrosis following cell, tissue, and organ hypothermic preservation. These studies have encompassed a variety of systems, including corneal and renal cells, hepatocytes, keratinocytes, liver tissue, and vascular cells and tissue.

Many of these reports have also described the development of next- generation hypothermic storage media, leading to improved cell storage in terms of both storage time and cell and tissue quality. Snyder et al., recently described the result of molecular investigations utilizing these new solutions in the preservation of cardiomyocytes. In this study, a triad of assessment regimes was employed to determine the preservation efficacy of several media including culture media, ViaSpan, and HypoThermosol-FRS. The study reported that through development of technologies based on the principles discussed by Taylor, as well as those of molecular control, cardiomyocyte hypothermic storage could be extended from one day at 4°C in media or ViaSpan to three days in HypoThermosol-FRS. Not only was cell viability maintained, but also mitochondrial activity and spontaneous contractile function of the cardiomyocytes was at levels near those of untreated controls.

Further, Mathew et al. recently detailed the combination of next-generation preservation solutions with TAC to effect even greater protection in liver and endothelial cells. Takesu et al. described the improved protection of hepatocytes by the addition of ascorbic acid-2 glucoside to ViaSpan. Numerous other studies have reported on the benefits of attempts to control stress, and therefore molecular responses, during hypothermic storage in fibroblasts, hepatocytes, and endothelial and corneal cells. These reports have shown that through molecular control of a cell's stress response to low temperature, preservation efficacy can be increased. However, as with cryopreservation, it has been shown that alterations in the preservation media formulation have had the most profound and beneficial effect on hypothermic storage efficacy.

From the discussion presented in this chapter it is clear that there has been a significant level of activity in the preservation sciences over the last several years aimed at molecular-based controls. With the identification of apoptosis and a cellular, molecular-based response to cryopreservation, our approaches to cryopreservation have changed significantly. No longer are researchers limited to preservation processes that primarily provide structural support following thawing. We have to look no further than stem cell research to see why this is such a critical and monumental milestone for the cryopreservation sciences. With molecular findings and the resulting technologies, systems that were thought to be "*uncryopreservable*" from a practical application prospective—such as cardiomyocytes, neurons, and neural progenitor cells, engineered cells and tissues, cell therapy products, etc.—can now be cryopreserved successfully.

Even in primary cell systems such as hepatocytes, successful cryopreservation has become a more attainable and acceptable option for cell storage due to the advances in the maintenance of cellular function following cryopreservation. This in itself is valuable, in that cryopreserved hepatocytes can now be more effectively utilized in basic science and drug discovery applications. With increases in post-thaw hepatocyte (as well as other cells) function and viability to levels near those of untreated controls, one can now avoid the necessity of "dual criteria" and "abstract" correlations of data sets as nonspecific evaluative tools.

Without a doubt, molecular-based study has propelled the cryobiological sciences to the forefront of scientific discovery. As our understanding continues to grow, advancements will continue to push the present-day limits of successful preservation. As such, given the current direction of the cryopreservation science, it appears that the future lies in the combination of next-generation solution-specific strategies TAC or GASP to improve post-thaw survival and failure.

6

Preservation of Cellular Therapies

Cell-based therapies continue to evolve and are becoming the standard of care for a wide range of human diseases and injuries. Most people are familiar with the transfusion of red blood cells to treat traumatic blood loss or disease-based anemia. Similarly, tens of thousands of people each year are treated with a bone marrow transplant. Cell-based therapies are based on the scientific and clinical platform created by blood component transfusions and hematopoietic stem cell transplantations. The purpose of this chapter is not to discuss clinical preservation protocols associated with blood components or traditional hematopoietic stem cell transplants; instead, the focus is on cell-based therapies in which the cells have undergone significant manipulation *ex vivo*. An excellent review of present clinical preservation practices for blood components and bone marrow can be found in a recent review by Sputtek and Sputtek. Specifically, the cells might be cultured to expand cell number, selected according to subpopulations, or subjected to genetic or biological modification. The manipulation of the cells outside of the body results in a product that differs greatly from that which is harvested from the body, and these differences influence the biological activity and biophysical properties of the cells and consequently the methods of preservation that should be used.

Overview of Cell-based Therapies

A wide range of cell-based therapies have been used to treat human disease. Islets of Langerhans are transplanted to treat diabetes, and cell and tissue transplants are used to treat Parkinson's disease and other neurological disorders. Isolated hepatocytes are also transplanted for the treatment of liver disease. Blood cell-based therapies have become the standard of care for some diseases, such as leukemia, and an even larger number of blood cell-based therapies are in preclinical development. Processing protocols for blood cell-based therapies are much more mature and extensive. The focus of this chapter is the preservation of bone marrow cell-based therapies.

Advances in our understanding of hematopoiesis and immunology fuel the development of a wide range of hematopoietic cell-based therapies. For example, adoptive immunotherapy typically involves removing lymphocytes from the patient, expanding cell numbers, and modifying the cells to enhance specific immune function. Clinical studies of lymphokine activated killer (LAK) cells have been performed for the treatment of advanced cancer. However, studies evaluating the effect of both LAK cell and tumor infiltrating lymphocyte (TIL) infusions for the treatment of cancer do not suggest that these therapies provide a significant clinical benefit.

In contrast, when donor lymphocyte infusions are used to treat patients who have relapsed after an allogeneic hematopoietic stem cell transplant, a clear clinical benefit has been found. Donor lymphocyte infusions (DLI) can be performed using minimally manipulated cells obtained using apheresis. Other forms of DLI involve the infusion of lymphocytes depleted of or enriched with specific cell subsets or cells transduced with suicide genes. Common clinical practice is to cryopreserve DLI using conventional cryopreservation protocols for hematopoietic stem cell products, and the therapy typically involves multiple infusions.

Early studies in lymphocyte-based adoptive immunotherapy led to the use of antigen-presenting cells (specifically dendritic cells) to stimulate lymphocyte response *in vivo*. Small numbers of activated dendritic cells are effective in generating an immune response against different pathogens (i.e., viruses and tumor cells). The majority of dendritic cell therapies involve taking monocytes or hematopoietic progenitor cells ($CD34^+$ cells) from a donor (autologous or allogenic), culturing the cells in cytokine-supplemented media to induce maturation, and exposing the cells to proteins from the patient's tumor cells, followed by reinfusion of the cells into a patient. Early clinical studies have shown that dendritic cell therapy is nontoxic and effective in certain patient groups. However, more controlled clinical studies are needed to prove efficacy.

Immunotherapy is also used to treat viral infections. For example, both lymphocytes and dendritic cells have been studied for the treatment of infection by the human immunodeficiency virus (HIV). Viral infection after hematopoietic stem cell transplantation is still a significant source of morbidity and mortality. Both cytomegalovirus (CMV) and Epstein–Barr infections have been treated with immunotherapy. As in the case of adoptive immunotherapy for the treatment of cancer, effectiveness of the therapy relies on the ability to cultivate and reinfuse virus-specific cells (lymphocytes or dendritic cells).

Hematopoietic stem cell transplants have become the standard of care for malignant and nonmalignant blood diseases. For instance, leukemia, aplastic anemia, or thalassemia can be treated with hematopoietic stem cell transplants. Hematopoietic stem cell rescue has also been used to restore bone marrow function to patients receiving high-dose chemotherapy or radiation therapy for the treatment of solid tumors. For almost a decade, hematopoietic stem cell transplants have been used to treat a wide range of autoimmune diseases such as rheumatoid arthritis, systemic lupus erythematosus, and multiple sclerosis.

Recent studies suggest that hematopoietic stem cell transplantation can be used for the treatment of a wide range of medical conditions. For example, early clinical studies suggest that hematopoietic stem cell transplantation can be used in the treatment of myocardial infarctions. The limited studies performed to date show an improvement in left ventricular function for patients receiving injections of hematopoietic progenitor cells or unfractionated bone marrow cells after myocardial infarction. The mechanism for the observed improvement in cardiac function is still unknown. Further studies are needed to elucidate the mechanisms and establish conclusively the safety and efficacy of this type of treatment.

Cells harvested from the bone marrow include both hematopoietic cells (stem cells and blood cells at different stages of maturation) and nonhematopoietic cells (adipocytes, endothelial cells, stromal fibroblasts, reticular cells, smooth muscle cells, etc.). More commonly known as mesenchymal stem cells or marrow stromal cells (MSC), these cells were discovered by Friedenstein and colleagues over thirty years ago. They established that these cells could differentiate into osteoblasts, adipocytes, and chondrocytes *in vitro*. Subsequent investigations have confirmed these observations and further demonstrated that the cells can differentiate in culture into muscle, neuronal precursors, and cardiomyocytes. Additionally, animal studies show potential effectiveness for MSC-based therapy for the treatment of ostengenesis imprefecta, Parkinsonism, spinal cord injury, stroke, myelin deficiency, and lung diseases. Early human clinical trials for the use of MSCs to treat osteogenesis imperfecta, Hurler's syndrome, and metachromatic leukodystrophy suggest that MSCs can be effective in treating these disorders as well.

With the exception of bone marrow transplantation for the treatment of leukemia and stem cell rescue, all the therapies described above require extensive *ex vivo* cell manipulation. The manipulation of the cell product (expansion, enrichment or depletion of specific subsets, and biological or genetic modification) results in a significant change in the composition of the final product from the initial harvested product. Also, the time required for some of these protocols ranges from hours to weeks. Furthermore, specific cell therapies may require multiple injections or infusions of cells over a time period. All of these factors influence the preservation protocol appropriate for the desired therapy.

As more and more cell-based therapies become the standard of care, the need to develop improved and appropriate preservation techniques will only become more urgent. A recent survey by Read and Sullivan shows that the most significant growth in products processed at clinical cell-processing centers involve what have been considered nontraditional therapies described previously in this chapter. The remaining sections in this chapter will describe both the scientific and clinical need for reengineering preservation protocols for cell-based therapies.

Biophysical Properties of *Ex Vivo* Cultured Cells

As described by Mazur in a recent review, "Cryobiology is not only concerned with the state of water in cells and the role of water in the structure and function of components but is also equally concerned with factors affecting the movement of water into and out of cells." To that end, the water content and water permeability for lymphocytes and hematopoietic progenitor cells have been measured as a function of temperature and time in culture.

Ex vivo culture of hematopoietic progenitor cells resulted in a significant change in cell size. The average cross-sectional area of freshly isolated $CD34^+$ cells (hematopoietic stem cells) from a normal donor was observed to be 57 ± 13 μm^2. After culture and transduction of the cells (a five- day protocol), the cross-sectional area of the cells had increased to 116 ± 33 μm^2, which was statistically greater than the freshly isolated cells ($p<0.0001$). The total volume of a cell consists of water available for transport, solids, and water that is not available for transport. The osmotically inactive cell volume fraction represents the relative contribution of these different components. For $CD34^+$ cells from normal donors, the osmotically inactive cell volume fraction was initially 0.29 and reduced to 0.20 after three days in culture. Combining calculations of cell volume and osmotically inactive cell volume fraction, the water content of the cells increased roughly twofold during the brief culture period.

We were interested not only in water content but water transport as well. The lymphocytes, which were cultured *ex vivo* for two weeks to permit expansion of total cell number and genetic modification, were frozen on a cryomicroscope. The volumetric response of the cells was observed during the freezing process, and the corresponding images were analyzed. The volumetric response of the lymphocytes that were cultured and transduced is compared to freshly isolated lymphocytes from a normal donor. The normalized cell volume (volume

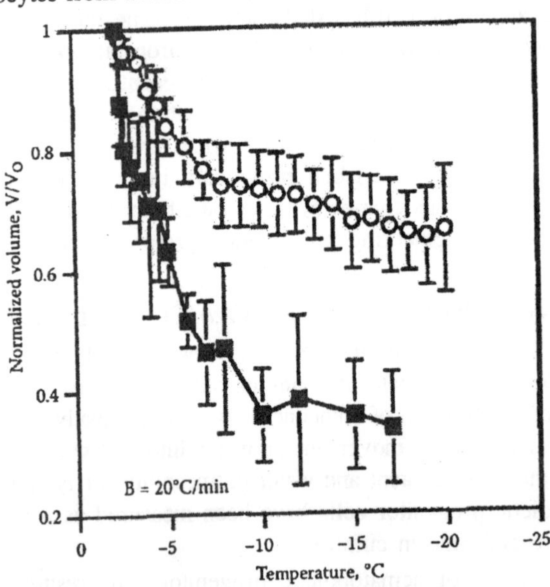

Fig. 6.1. Normalized cell volume as a function of temperature for cultured and transduced lymphocytes from a donor with MPS II and freshly isolatd mononuclear cells from a normal donor.

at a given temperature divided by the initial volume) decreased more rapidly as a function of temperature for the cultured cells than the fresh cells from a normal donor. In addition, the minimum normalized volume achieved by the cultured cells was significantly less than that typically observed for the fresh cells.

The efflux of water that results in the observed change in cell volume can be analyzed using a nonequilibrium model developed by Mazur. Assuming that equilibria of temperature and pressure prevail between the intra- and extracellular media, the following equation can be written:

$$\frac{dV}{dT} = \frac{L_p ART}{v_w B}\left[\ln\left(\frac{V - V_b}{(V - V_b) + v_w(v_s n_s)}\right) - \frac{\Delta H_f}{R}\left(\frac{1}{T_r} - \frac{1}{T}\right)\right] \quad \text{...(1)}$$

in which V is the cell volume, T is the temperature (absolute), L_p is the hydraulic permeability, A is the surface area of the cell, B is the cooling rate, v_w is the partial molar volume of water, V_b is the osmotically inactive cell volume fraction, v_s is the the partial molar volume of salt, n_s dissociation constant for NaCl, ΔH_f is the latent heat of fusion of water, R is the Universal Gas Constant, and T_r is the equilibrium freezing temperature for pure water.

The hydraulic permeability of the membrane, L_p, is a function of the temperature. Assuming an Arrhenius relationship, the permeability is expressed as a function of temperature:

$$L_p = L_{pg} \exp\left(-\frac{E_{lp}}{R}\left(\frac{1}{T_r} - \frac{1}{T}\right)\right) \quad \text{...(2)}$$

in which L_{pg} is the permeability of the cell membrane to water at the reference temperature T_R, and E_{lp} is the apparent activation energy for the water transport process. Analysis of the volumetric data obtained from cryomicroscopy experiments using Equations 1 and 2 indicate that the average values of the water permeability parameters, L_{pg} and E_{lp}, for the cultured and transduced cells from a donor with mucopolysaccharidosis Type II (MPS II) were determined to be 5.3 × 10^{-14} m^3/Ns and 156 kJ/mol, respectively. For cells from a normal donor, L_{pg} and E_{lp} were 1.4 × 10^{-14} m^3/Ns and 168 kJ/mol, respectively. Thus, the *ex vivo* culture process resulted in an increase in both L_{pg} and E_{lp} cells for the cells. Similar studies were performed using $CD34^+$ (hematopoietic progenitor cells) and, as with the lymphocyte studies, significant changes in L_{pg} and E_{lp} were observed between fresh and cultured cells. Cultured cells from a normal donor exhibited a decrease in both L_{pg} and E_{lp} with time in culture. This shift in permeability parameters differs from those observed with lymphocytes and may reflect differences in the basic biology between mature and progenitor cells.

Unfortunately, these studies do not suggest a specific mechanism for the change in water content or water transport properties of the cells; however, previous studies of the freezing behavior of hepatocytes may suggest a potential mechanism. Increases in post-thaw viability and function were observed for hepatocytes cryopreserved after preculture in the form of aggregates. Studies using NMR showed distinct shifts in membrane composition and shifts in metabolic pathways between freshly isolated and cultured hepatocytes. These shifts in membrane composition and metabolic activity occurred over a relatively short period of time (24 h culture), and the shifts in membrane composition (increase in cholesterol content) were consistent with the change in water permeability measured. These studies suggest that differences in the environment (nutrient and oxygen concentrations, shear forces, cell-to-cell contact, etc.) between *in vitro* and *in vivo* culture may influence cell metabolism for hematopoietic cells cultured *ex vivo*. This influence may result in the changes in water content and membrane permeability observed. Additional studies are needed, however, to conclusively establish this link.

Stem cells capable of reconstituting hematopoiesis can be obtained from various sources — peripheral blood, bone marrow, umbilical cord blood, and fetal liver cells — and they share certain traits such as the expression of cell surface marker $CD34^+$ and the ability to reconstitute hematopoiesis in humans. The different sources of hematopoietic stem cells reflect different ontogenies. Distinct changes in proliferative and differentiation potential for the hematopoietic stem cell compartment have been observed throughout development and life. Investigators have examined the water content and permeability characteristics of hematopoietic stem cells from these different sources, and their studies provide insight into the role of ontogeny on the freezing behavior of a cell-based therapy.

Three different biophysical properties — osmotically inactive cell volume fraction, water permeability, and permeability to dimethly sulfoxide (Me_2SO). Distinct differences in the osmotically inactive cell volume fraction can be observed. Human fetal liver-derived hematopoietic cells exhibit an osmotically inactive cell volume fraction of approximately 48% with decreasing values for V_b with maturation (in terms of ontogeny). Peripheral blood stem cells (PBSCs) are the exception to this trend with a *Vb* of approximately 29%, and that difference may reflect the influence of the use of a drug (Granulocyte-Colony Stimulating Factor) to stimulate migration of stem cells into the peripheral blood. The difference between the permeability of the cells to water or Me_2SO for the different cell types studied was insignificant. Variations in biophysical characteristics for cells with ontogeny have distinct implications for the development of cryopreservation protocols for cells and tissues derived from adult or embryonic stem cells. As stem-cell-based products are brought to clinical and commercial use, the cryopreservation protocols may need to reflect the distinct differences in the cells resulting from *ex vivo* culture and/or ontogeny.

POST-THAW SURVIVAL

The changes in biophysical properties with time in culture described in the previous section are only one descriptor of the freezing response. We are also interested in the post-thaw survival of cells that are cultured, frozen, and thawed. One study measured the post-thaw viability of lymphocytes frozen immediately after isolation and those frozen after culture and transduction. For cells from the same donor, the post-thaw viability of lymphocytes significantly decreased from 87% for freshly isolated cells to 69% for the cultured and transduced cells ($p<0.0001$). The culture and transduction protocol for the lymphocytes was approximately two weeks in duration. These findings suggest that long-term culture of the cells may alter their ability to survive the stresses of freezing and thawing. Yokomuro and colleagues observed similar results in a study of cryopreserved cardiomyocytes. They observed that the post-thaw functional recovery of the cryopreserved cardiomyocytes was reduced with increased cell passage. Thus, the ability to cryopreserve primary cells versus immortalized cell lines may be influenced by culture period.

The previous paragraph describes a situation in which *in vitro* culture resulted in a decrease in post-thaw survival. We were also interested in determining whether shorter periods of culture had the same effect. To that end, peripheral blood lymphocytes were cultured using the same basic protocol, except that the cells were not transduced with a retroviral vector and the entire culture protocol was limited to one week in duration. Samples of the cells were removed from culture and cryopreserved using the same protocol (10% Me_2SO, 1°C/min) at day zero, three, and seven in culture. Post-thaw cell recovery — the number of viable cells post-thaw divided by the number of viable cells

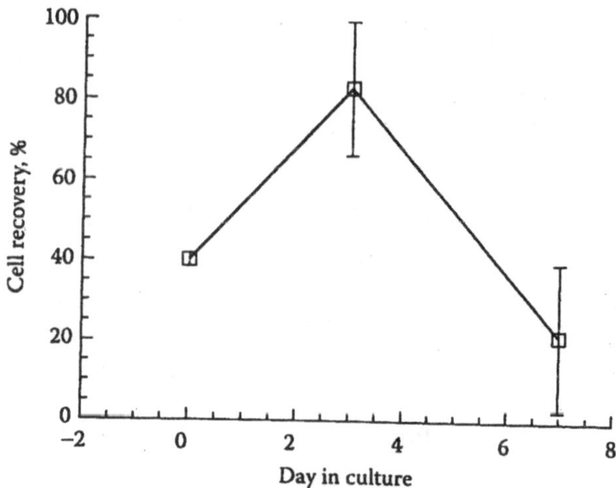

Fig. 6.2. Cell recovery as a function of day in culture of peripheral blood lymphocytes frozen preserved in 10% DMSO at 1°C/min.

prefreeze — was determined 48 hours post-thaw. The cell recovery increased from 40.5 ± 2.5% at day zero to 84 ± 18% at day three in culture. After seven days in culture, the post-thaw cell recovery was 22 ± 18%. The cell recovery at day zero was not statistically different than day seven, but the cell recovery at day three was distinctly greater than both days zero and seven ($p<0.0001$). Although this study does not suggest a specific mechanism for the observed differences in post-thaw recovery with time in culture, lymphocyte activation and death in culture have been analyzed in studies too numerous to list. Medema and Borst summarized many of these studies in a review. They observed that after activation, primary T-lymphocytes are resistant to activation-induced cell death for the first few days in culture. However, these cells lose this resistance to apoptosis after three days in culture. The culture protocol for the lymphocytes included activation of the cells by exposure to anti-CD3 antibody at the beginning of the protocol. Thus, the observed increase in post-thaw recovery at three days in culture may reflect the influence of cell resistance to apoptosis. Further studies will be needed to confirm this hypothesis.

Post-thaw survival can also vary with time post-thaw. The total number of frozen and thawed lymphocytes in culture is determined as a function of time post-thaw. The number of viable cells in the culture bag declines over a 48- to 72-hour period of time and finally increases as proliferation exceeds cell death. Baust and colleagues also observed this decline in viability with time in culture post-thaw with MDCK

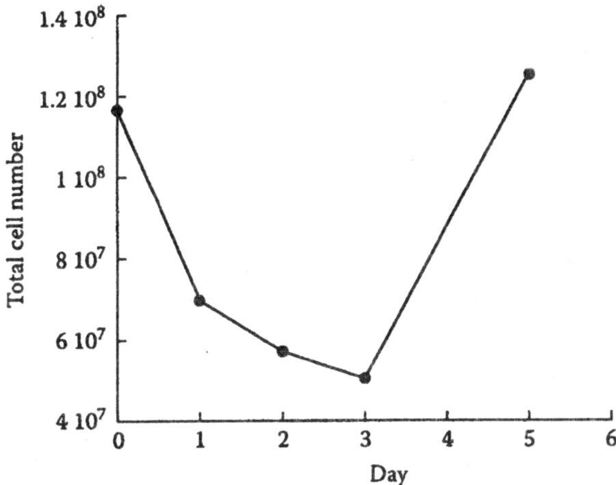

Fig. 6.3. Growth in peripheral blood lymphocytes (PBLs) from a patient with monopolysaccharidosis type II after cryopreservation.

cells; they went on to establish a link between post-thaw apoptosis and the decline of post-thaw survival.

Several investigators have documented and studied post-thaw apoptosis for hematopoietic cells. DeBoer and colleagues observed that a significant fraction of hematopoietic stem cells ($CD34^+$ cells) expressed early markers of apoptosis. If corrected for the fraction of cells expressing early stage apoptosis, the post-thaw viability of $CD34^+$ cells decreased from 78% (measured with viability stains alone) to 42% (measured with apoptosis and viability stains combined). The authors further asserted that post-thaw apoptosis may influence the dose of $CD34^+$ cells required for engraftment. Stroh and colleagues studied the specific mechanism for the observed post-thaw apoptosis and found that both primary hematopoietic cells and hematopoietic cell lines exhibited post-thaw apoptosis mediated by the mitochondria and activated specifically by caspase 3. They observed that only overexpression of Bcl-2 resulted in protection of the cells from post-thaw apoptosis and that the use of caspase inhibitors conferred protection from post-thaw apoptosis. Similar results were obtained by Sarkar and colleagues studying the post-thaw apoptosis of T-lymphocytes. Sarkar and colleagues also investigated the potential for using caspase inhibition to prevent apoptosis or cytokines to rescue cells from post-thaw apoptosis. Post-thaw apoptosis has been shown to influence the viability of hepatocytes. As with hematopoietic cells, post-thaw apoptosis for hepatocytes appears to be mediated by the mitochondria and activated by caspases. Cell losses resulting from apoptosis are typically observed between 24 and 72 hours post-thaw depending upon cell type. A limited number of studies have examined the role of freezing and thawing on longer-term cell survival. Frozen and thawed retinal pigment epithelial cells exhibited altered cell membrane protein structure and enzymatic function as well as damage to the DNA as a result of freezing. A subsequent study of these cells by Honda and colleagues demonstrated that cryopreservation decreased proliferation of the cells post-thaw and shortened telomeres. These studies suggest that molecular-level phenomena may play an important role in both short- and long-term freezing response.

Clinical Issues

Effective cryopreservation of cell-based therapies is essential to their clinical application. Cryopreservation permits coordination of culture protocols with patient care regimes, completion of safety and quality control testing before administration, and transportation of the product from the processing facility to the site of use (doctor's office, hospital, etc.). Specific challenges arise in the application of cryopreservation to a clinical context. For example, clinical application of cryopreservation must involve the use of solutions and reagents appropriate for human use.

The reinfusion of nonfrozen (fresh) hematopoietic cells is typically not associated with adverse reactions, in contrast, the infusion of cryopreserved hematopoietic stem cells (bone marrow, PBSC, and cord blood) is commonly associated with adverse events. Studies suggest that one source of adverse reactions results from the infusion of cell fragments. Cell losses from the stresses of freezing and thawing can be significant (>30%), and most of the dead cells are no longer intact with cell membranes disrupted and intracellular contents released. The infusion of cell fragments has been associated with fevers, chills, hypotension, and respiratory distress. The severity of the reaction is a function of the dose of fragments infused.

Controlled clinical studies show that Me_2SO and not cell fragments is responsible for the most severe adverse reactions. Nearly all patients receiving cryopreserved hematopoietic stem cell transplants containing Me_2SO experience nausea, chills, hypotension, dyspnea, and cardiac arrythmia. In a smaller number of patients receiving infusions with hematopoietic stem cell products, case studies also illustrate more serious reactions including cardiac arrest, transient heart blockage, neurological toxicity, renal failure, and respiratory arrest. Studies of hematopoietic stem cell transplants in children show that reactions in this patient group are more severe. Children generally receive a greater dose of Me_2SO per kg of body weight for a transplant than adults, and consequently, children experience both more frequent and more severe adverse reactions. Several of these studies that examined the relationship between transfusion reactions and Me_2SO content of the transplanted cells noted either that the greater the dose of Me_2SO, the more severe the adverse reaction, or that removing Me_2SO reduced the incidence and severity of the transfusion reaction.

Based on the clinical studies documenting the severity of reactions for children, transplantation protocols for umbilical cord blood specify post-thaw washing of the cells to remove Me_2SO. Unfortunately, washing cells (by centrifugation or automated cell washers) prior to infusion to remove Me_2SO results in significant cell losses. Antonenas and colleagues observed a loss of 27 to 30% of nucleated cells resulting from post-thaw washing of cord blood. A more recent study by Perotti and colleagues observed a similar loss in nucleated cell counts for cord blood units washed with an automated cell washer. Cell losses, in particular for cord blood transplants, have a significant influence on transplant outcome. Specifically, reduced cell numbers are associated with longer periods of engraftment and increased morbidity and mortality. Therein lies the conundrum: direct infusion of cryopreserved hematopoietic stem cells results in significant adverse reactions, but washing cells to remove Me_2SO results in significant cell losses with the risk of increased patient morbidity and mortality.

Cryopreservation solutions are composed of tissue culture medium, cryoprotective agent (typically Me_2SO), and animal serum (typically fetal calf serum (FCS)). As described, Me_2SO is associated with adverse reactions, as is the use of animal sera in culture media for human cell therapy. Other components of cryopreservation solutions may also be inappropriate for human therapeutic application. Selvaggi and colleagues documented a range of unfavorable reactions resulting from the *ex vivo* culture of lymphocytes using FCS-containing media. Patients receiving cell infusions exhibited hypotension, rigors, arthralgias, myalgias, headache, and/or malaise. Furthermore, the majority of patients formed antibodies to specific components in FCS within two cell infusions. This finding echoed studies performed over two decades earlier by Irie and colleagues. Both studies suggest that the use of animal sera in cryopreservation solutions for human therapeutic applications is not appropriate. Fortunately, the majority of clinical cryopreservation protocols do not use animal sera in the cryopreservation solution but rather autologous serum/plasma or human serum albumin (HSA). The quality of the serum/plasma can vary greatly based on the health status of the patient. Consequently, it may be difficult to obtain a consistent outcome for the cryopreservation protocol when autologous serum is used. HSA has been adopted as an alternative for autologous serum, but concerns over supply and the transmission of adventitious agents makes HSA a less-than-ideal substitute. The mechanism by which proteins act to protect cells during freezing is as yet unknown; as a consequence, little has been done to develop alternatives.

By volume the principal component of clinical cryopreservation solutions typically consists of tissue culture media. Tissue culture media can easily contain more than fifty components, including amino acids, inorganic salts, vitamins, and trace elements. Many of these components are not available in therapeutic grade, and the safety of these solutions for humans has not been studied. A limited number of investigators have examined the use of plasma expanders or electrolyte solutions as a base for cryopreservation solutions or as a short term storage solution for hematopoietic stem cells. These studies suggest that infusible grade solutions could be substituted for tissue culture media in both cryopreservation and liquid storage solutions. Little movement, however, has been made to standardize the use of licensed electrolyte solutions or plasma expanders as a basis for clinical cryopreservation solutions.

Future Directions

The preservation of cell-based therapies presents a unique opportunity and challenge. Studies demonstrate that *ex vivo* culture can influence cell metabolism and membrane composition, which in turn can influence post-thaw survival. Future studies elucidating the cellular and molecular mechanisms that influence cell survival (beyond apoptosis) could be important in advancing our basic understanding of cellular or molecular response to the stresses of freezing and thawing and lead to the development of a novel class of cryoprotective agents ("*molecular modifiers*") that exploit this understanding to enhance post-thaw survival. We may also be able to develop *ex vivo* culture protocols that minimize detrimental cell changes (increases in membrane cholesterol content, telomere shortening, and shifts in metabolism to pathways typically associated with stress response) and therefore maximize post-thaw recovery.

At the present time, stem cell-based therapies have typically involved adult stem cells. The preservation of embryonic stem cell-based therapies is an important emerging challenge. Early studies have shown the ability to cryopreserve embryonic stem cells and the ability to preserve neuronal progenitor cells derived from embryonic stem cells. Our difficulty in preserving certain mature cell types (e.g., *granulocytes* and *hepatocytes*) suggests that post-thaw survival may vary with differentiation. Our understanding of biophysical (water transport properties) and molecular properties on the post-thaw survival of cells may be essential to the rational design of preservation protocols for embryonic stem cell-based therapies.

Worldwide, a considerable financial and human resource investment has already been made in banking hematopoietic stem cells. The hundreds of thousands of hematopoietic stem cell product (typically cord blood) presently stored in clinical banks use conventional cryopreservation solutions containing Me_2SO. The development of improved methods of post-thaw processing to minimize infusion of Me_2SO and cell losses would improve the safety and efficacy of these products.

Finally, cell-based therapies developed for human therapeutic application require the development of preservation methods appropriate for human use. In particular, current cryopreservation solutions contain components associated with adverse reactions or are not available in therapeutic grade. Alternatives for tissue culture media and animal sera have been developed and partially adopted but are not yet mainstream. The development of Me_2SO-free cryopreservation solutions remains a significant hurdle, but it would represent a huge step forward in improving the safety of cryopreserved cell products. Further work is needed to develop solutions safe for human use and obtain regulatory approval.

7

Biopreservation of Tissues and Organs

Transplantation science calls for effective methods of preservation since it is unavoidable that donor cells, tissues, and organs are required to withstand a period of ischemia and hypoxia as part of any transplantation protocol. Historically, interest in isolated organ preservation was recorded at the beginning of the present century when Carrel was the first to explore techniques for preserving tissues for the purposes of transplantation. Carrel was awarded the Nobel Prize in 1912 and in subsequent studies in the 1930s investigating normothermic organ perfusion with serum or synthetic perfusates, he laid down the basic requirements for artificial perfusion technology relating to organ preservation and cardiopulmonary bypass. It was recognized at that time that the viability of perfused tissues depends on a variety of factors that remain pertinent today. These included the need for the perfusion fluid (blood or synthetic solutions) to be free of emboli (gaseous or particulate), including agglutinated corpuscles if blood was used. Physical and chemical characteristics of the perfusion medium were also recognized to be of crucial importance. These include temperature, osmotic pressure, pH, oxygenation, and chemical composition.

During the next two decades perfusion techniques were improved and invariably combined with hypothermia for effective preservation. While the protective effects of low temperatures have been known and explored by mankind since the dawn of civilization and began to be documented as early as the seventeenth century, the scientific basis for cell death following ischemia and its amelioration by hypothermia has only begun to be understood during the past fifty years. The modern era of low temperature cell preservation, which began in the 1950s, involves both hypothermic preservation at temperatures above 0°C and also cryopreservation at subzero temperatures utilizing freezing or vitrification techniques. The fundamental basis of low-temperature preservation is to use cold as a physical means of depressing function in a reversible way, i.e., to achieve a state of "*suspended animation.*" The principles and mechanistic basis of cryopreservation that currently permit a wide variety of single cells and some simple tissues to be stored indefinitely at deep subzero temperatures will be discussed in other chapters. At present, most complex multicellular tissues and organs cannot be cryopreserved without incurring intolerable levels of cryoinjury and effective methods of preservation rely upon hypothermic storage at temperatures above the freezing point. The purpose of this chapter is to summarize the basic principles upon which mammalian cells can be safely held in suspended animation in the cold.

The cells of poikilothermic animals have adapted through evolutionary processes to survival in the cold, but for the cells of euthermic animals, including man, cold is a stress that may be tolerated depending upon the degree and duration. Some mammals have adapted to these cold stresses through the process of hibernation, and emerging clues for the mechanisms involved may, in the future, provide benefits for interventional strategies of hypothermic preservation of tissues harvested from nonhibernators. What then is the scientific basis for the longstanding use of hypothermia as the cornerstone of virtually all effective methods of multicellular tissue and organ preservation? In this chapter, we will outline the basis for cell survival in the cold by considering the cell in relation to its environment, the known effects of ischemia and anoxia, the influence of hypothermia on ischemic events, and, finally, the strategies of interventional control of the extracellular environment to optimize preservation.

Cell in Relation to its Environment

Differences between the Extracellular Environment and the Intracellular milieu, and their Control

A cell is defined morphologically by its limiting envelope, the plasma membrane, which separates it from its neighbors in a tissue or from its fluid environment. Within the body, the extracellular fluid bathing cells is maintained within closely regulated limits by a variety of control mechanisms and cells maintain a different, but even more constant intracellular milieu. The contrasting balance of ions between the intracellular and extracellular spaces of a typical mammalian cell, with the extracellular complement of ions representing the typical composition of a balanced salt solution such as Ringer's or Krebs' etc. buffer solutions for *in vitro* cell maintenance.

Cells are able to maintain a stable internal environment very different from that which surrounds them by means of their membrane pumps, and principally by the sodium pump mediated by Na^+ - K^+ ATPase. As illustrated, this is an energy-consuming process that extrudes sodium and accumulates potassium in the ratio of 3:2 (i.e., the pump is electrogenic and contributes to the cell membrane electrical potential). Each ion tends to diffuse back down its concentration gradient, but because potassium diffuses out more rapidly than sodium can diffuse in, the net loss of cations from the cell charges the cell membrane (negative internally). The voltage gradient also influences the distribution of anions, and Cl^-, which diffuses freely in response to the membrane potential, is excluded. Therefore, the sodium pump in effect extrudes NaCl and, because the membrane is highly permeable to water, an osmotic quota of water (180 molecules/solute molecule) leaves the cell with the Na^+ and Cl^-. Thus, the sodium pump controls cell volume as well as ion distribution and the maintenance of the intracellular environment is dependent upon metabolic generation of high-energy phosphates.

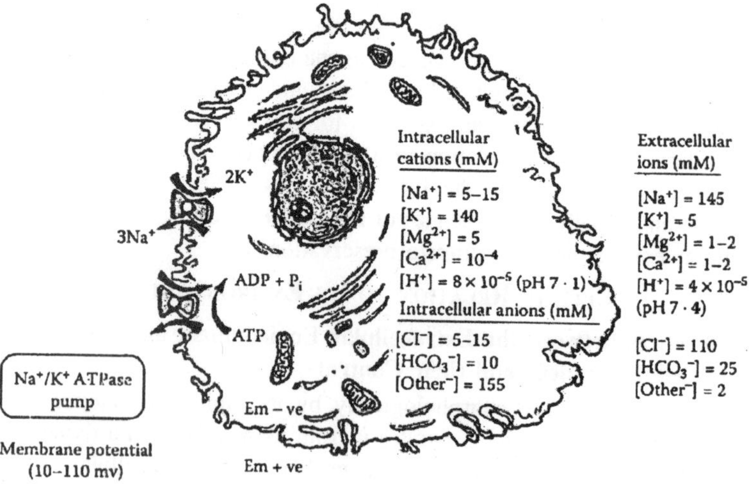

Fig. 7.1. Schematic diagram of a typical mammalian cell showing the contrasting balance of ions between the intracellular and extracellular compartments.

Apart from small inorganic ions, the cell contains a variety of small organic molecules and metabolites as well as macromolecules that, unlike the inorganic ions, are not free to traverse the cell membrane, but are confined within the intracellular compartment. This has important implications for the control of cellular hydration and volume. Osmolality is a colligative property and as such, depends upon the actual number of particles present in solution rather than on the size or nature of the molecules. Hence, intracellular macromolecules themselves contribute very little to the osmolality of the cell interior since they are present in relatively small numbers compared with the number of small molecules and ions. However, macromolecules are invariably highly charged, attracting a large number of small counter-ions that contribute significantly to intracellular osmolality. In addition, a cell contains a high concentration of small organic molecules as

A. Sources of osmolality

Extracellular

Intracellular

Macro molecules

Small organic molecules and metabolites

Small inorganic ions

B. Control osmotic swelling

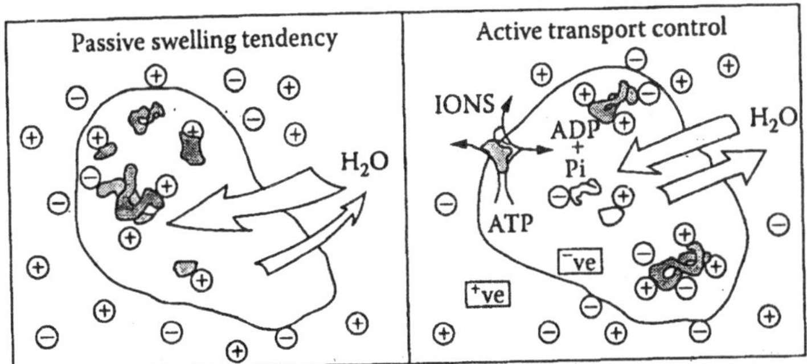

Fig. 7.2. The cell in relation to its environment: schematic representation of the sources of omolality inside and outside of cells and the role of active transport in controlling cellular hydration and cell volume.

products of metabolic processes and these impermeable metabolites are also associated with a complement of counter-ions.

In contrast, the osmolality of the extracellular fluid is due principally to small inorganic ions that leak slowly across the plasma membrane. The distribution of these ions is governed both by the presence of the impermeant charged molecules inside the cell and by the active ion pumps associated with the plasma membrane. These factors give rise to a greater concentration of inorganic ions inside the cell than outside at equilibrium, a phenomenon known as the Donnan effect.

Osmotic cell swelling

In essence, if the active ion transport processes are inhibited as a consequence of either energy deprivation and/or temperature reduction, then the impermeant solutes and colloidal material inside the cell give rise to water uptake. This in turn reduces the concentration of salt below that of the extracellular fluid and more salt penetrates making the cell hyperosmolar, thus inducing more and more water uptake until the cell eventually bursts, an event known as membrane rupture.

This colloid osmotic swelling is a very significant factor in the cell's relationship to its environment since the semipermeability of cell membranes vis-a-vis ions is only effectively maintained by metabolic processes that pump ions in or out of the cell. Where these processes fail the effective semipermeability is lost and the result is colloid osmotic swelling and lysis. Obviously, this is a very important phenomenon that has to be taken into consideration when designing methods of hypothermic cell preservation as discussed below.

Similar considerations apply to the composition of interstitial fluid surrounding cells and the plasma within capillaries. Because of the low permeability of capillaries to proteins, the interstitial fluid normally has a low concentration of these plasma constituents, which are predominantly serum albumin (67%; MW = 61,000) and serum globulins (30%) with much higher molecular mass (~100,000 Daltons). The interstitial fluid can be considered as an ultrafiltrate of plasma with the Gibbs-Donnan equilibrium applying to the distribution of ions.

The circulation is therefore responsible for maintaining a remarkably constant composition of the extracellular fluid with the osmolality being controlled at about 300 m osmol/kg to equate with that of the cytosol within cells as described above: cells will tolerate a perturbation of osmolality of only 10% without harm.

Essential Maintenance Processes

Under normothermic physiological conditions, control of a cell's peculiar internal environment with respect to its surroundings is a function of both the properties of the plasma membrane and of the cell's energy processes within the cell. The membranes are bilayer lipid-rich structures with associated embedded proteins that invariably function as transmembrane transporters of ions and small molecules, e.g., via specific channels. It is mentioned above that the plasma membrane provides a selective barrier to the diffusion of ions and other solutes, the distribution of which is regulated by active metabolic pumps that require a constant supply of energy. Adenosine triphosphate is the principal molecular store of chemical energy in the cell, and the biochemical pathways involved in the catabolism of nutrient substrates into a form of energy useful for driving biosynthetic reactions and other energy-requiring processes — such as control of the intracellular milieu — are well known and described in any standard biochemistry text.

Figure 7.3 shows a simplified outline of the three principal stages of catabolism that lead from basic nutrients through progressive oxidations to waste products with the yield of large quantities of ATP. The three major pathways involved in the oxidation of glucose are glycolysis, the citric acid cycle (Krebs cycle) and the electron transport chain. The glycolytic pathway is the only process by which cells can obtain energy in the absence of oxygen (anaerobically) and although of limited capacity this pathway is important for helping to sustain homeostasis in ischemic/anoxic cells.

Essential Role of the Circulation

In summarizing the fundamental processes of normal cellular function and homeostasis as a prelude to discussing the nature of ischemic injury and its prevention, it is important to emphasize the essential role of the circulation. The foregoing discussion has outlined in elementary terms the well-known basics of how a cell is sustained for normal function and the composition of the regulation of the interstitial fluid is an important function of the circulation. Other essential functions include the provision of an internal heat transfer system that closely regulates the temperature of tissues and organs. The circulation transports metabolites of which the fuels for respiration (glucose, fatty acids, and ketone bodies) are most important in the short term. For most tissues, respiration is predominantly aerobic with high demands for oxygen (e.g., 300 mL O_2/min/g dry weight of kidney

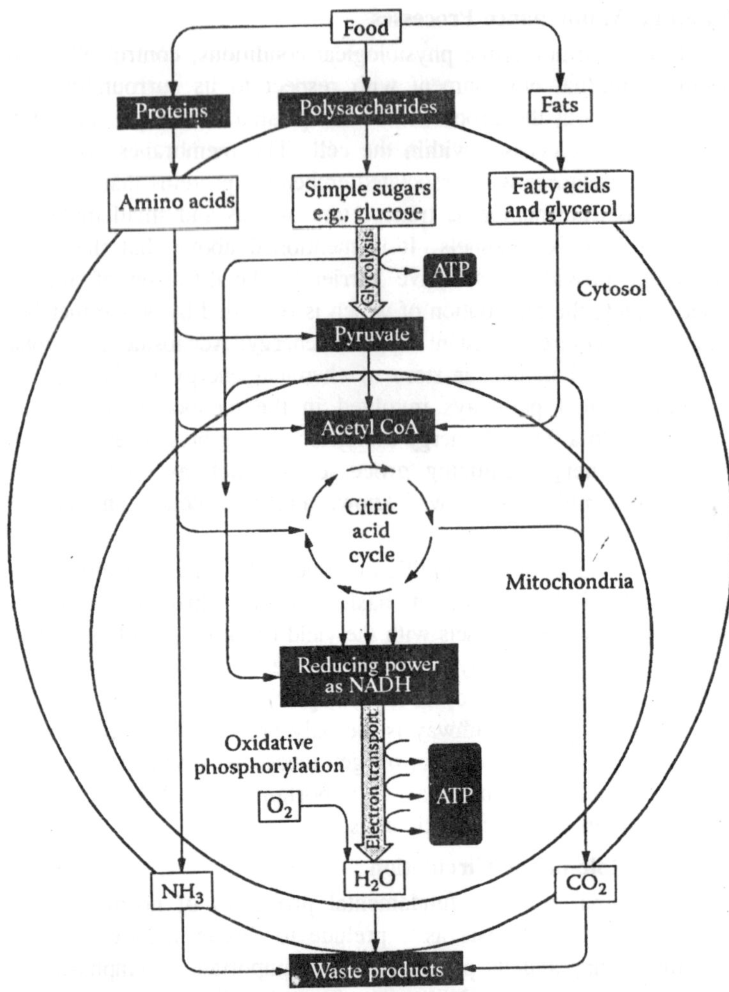

Fig. 7.3. A simplified diagram showing the principal stages of catabolism that produces ATP used to drive biosynthetic reactions and other energy-requiring processes in the cell such as control of the intracellular milieu.

renal cortex). The solubility of oxygen in tissue fluid is only about 24 mL/kg such that inhibition of aerobic energy production is very rapid following the onset of ischemia. Other necessary metabolites, besides those required for energy production, including amino acids, vitamins, glutathione, coenzymes, and many other factors, are also supplied to cells via the circulation. The catabolic products of cellular metabolism are removed by the vascular system. In light of the crucial role of the circulation in sustaining the normal environment and metabolic needs of cells, it is possible to appreciate the devastating consequences of a disruption to the vascular supply of a tissue or organ. A brief consideration will now be given to the effects of ischemia as a preface to a discussion of the positive and negative aspects of hypothermia as a protective modality against ischemic injury.

Synopsis of Ischemic and Hypoxic Injury

Interest in the pathophysiology of ischemia is not limited to just the transplantation community, since an understanding of the detrimental effects of a reduced blood supply to tissues is also of crucial importance in the treatment of cardiovascular diseases and neuropathological complications of acute cerebral vascular occlusions, clinically considered stroke. Consequently, there now exists a vast literature on the subject of ischemia and reperfusion injury and, although not yet complete, a considerable understanding of the nature of ischemic injury has emerged. This can only be outlined here in basic terms, but more complete accounts of the phenomena and mechanisms of ischemic injury in relation to the isolation of organs for transplantation have been published in the past, and a selection is cited here for reference. More recent accounts that focus on cellular and molecular mechanisms are to be found in the literature relating to ischemia/reperfusion injury in the heart and tissues of the central nervous system that are most sensitive to an ischemic insult.

Ischemic Cascade

Excision of a tissue for transplantation means that ischemia is total and inevitable even though the period may be brief. An immediate consequence of cessation of blood supply to an organ is deprivation of the supply of oxygen to the tissues, but anoxia (total deprivation) or hypoxia (partial deprivation) is only one of the many consequences of a lack of blood supply. The pivotal event is ATP depletion, which occurs within the first few minutes of oxygen deprivation. This early event leads immediately to a shift from aerobic to anaerobic metabolism (Pasteur effect), which very quickly becomes self-limiting with the production of lactate and protons. Cell depolarization also occurs very early in the cascade leading to a breakdown of ion homeostasis and a concatenation of other intracellular and membrane-associated events that eventually culminate in necrosis and cell death. A rise in the intracellular concentration of protons and calcium is at the center of many of the mechanisms now recognized to be contributory to cell death as a result of ischemia.

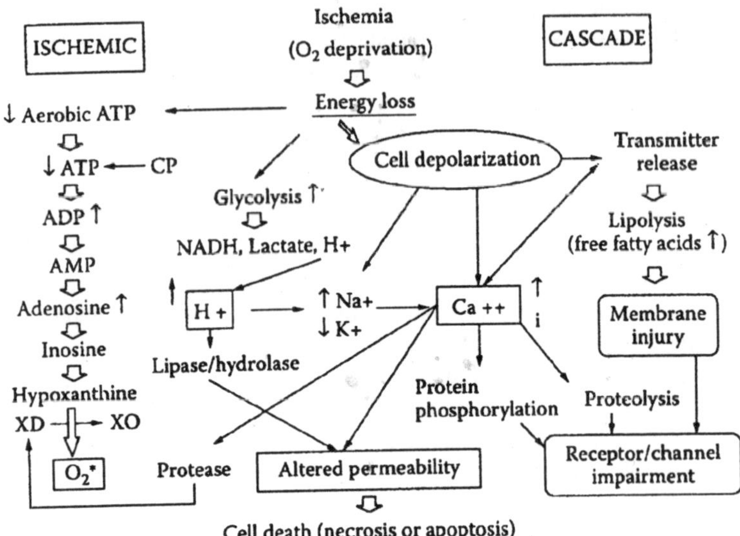

Fig. 7.4. A scheme for the principal metabolic and ionic changes that proceed after the initiation of ischemia.

It is well known that different cells have different susceptibilities to ischemic injury and this is due largely to a higher metabolic activity in the more sensitive tissues requiring a greater ongoing production of ATP. The progression from early onset reversible changes to subsequent irreversible injury is both time and temperature dependent. Moreover, a precise distinction between reversible and irreversible injury is difficult to specify. While some intermediate events such as ionic shifts may or may not be readily reversible, structural alterations in organelles, especially the mitochondria, are generally regarded as early signs of irreversible damage and the rupture of cell membranes is unequivocally fatal for the cell. Reduced temperatures can delay the progression towards a state beyond which cells are unable to recover normal function, and this is the basis of hypothermic preservation as discussed below. Under warm ischemic conditions, the time available during which cells will sustain only reversible changes is very restrictive for clinical procedures. For example, those tissues most sensitive to ischemia (e.g., heart and brain) are irreversibly damaged within a few minutes such that hypothermia is often needed as an adjunct to protect these organs *in vivo* during complex surgeries requiring lengthy periods of circulatory arrest. The ischemic tolerance of the brain is known to be only 6 minutes at 37°C, but is extended to nearly 60 minutes when body temperature is reduced to 17°C. During global warm cerebral ischemia, the high-energy phosphate stores of creatine phosphate are depleted within one minute; glucose and glycogen stores within 4 minutes; and ATP reserves within 5 to 7 minutes. In contrast, some transplantable visceral organs such as the kidneys are able to tolerate much longer periods of warm ischemia. Detrimental changes occur first and become most severe in the proximal convoluted tubules. Gross ischemic changes in some tubules are observed after 30 minutes, with total necrosis of the majority of tubules after 60 minutes. Nevertheless, 90% of experimental animals survived with kidneys that suffered 30 minutes of warm ischemia and 75% survived after 60 minutes of ischemia, showing that much of the damage is recoverable despite evidence of permanent histological injury.

The heart is highly intolerant of ischemia and continues to challenge researchers to devise ways of extending and improving methods of myocardial preservation. The heart is an obligate aerobic organ and the myocardium is exquisitely sensitive and dependent upon a continuous and adequate supply of oxygen for maintenance of normal contractile function. For the purposes of this discussion, it is appropriate to focus on the heart to summarize the effects of ischemia, which must be alleviated in order to prolong preservation.

Under aerobic conditions, mitochondrial oxidative phosphorylation provides the primary energy source for the myocardium, which is able to use a variety of substrates such as glucose, free fatty acids, lactate, pyruvate, acetate, ketone bodies, and amino acids. Myocardial oxygen reserve is exhausted within 8 seconds following the onset of global ischemia, and aerobic production of ATP ceases as the tissue oxygen tension falls below 5 torr. During the first phase of ischemia, the myocardium depends upon energy production from glycogenolysis and anaerobic glycolysis. However, this pathway is an intrinsically inefficient way of maintaining myocardial ATP and is ultimately inhibited during prolonged ischemia by the accumulation of glycolytic metabolic degradation products (NADH, lactate and H^+). ATP stores are temporarily buffered by the pool of creatine phosphate (CP), but this declines rapidly and is essentially depleted within 20 minutes of the onset of global ischemia. During the initial 5–10 minutes of ischemia, ATP levels do not drop significantly, but decline to 50–60% of pre-ischemic levels during 30 minutes of interrupted coronary circulation.

An important consequence of the increased glycolysis and ATP hydrolysis is the accumulation of protons in the cytoplasm, giving rise to a progressive development of intracellular acidosis. Apart from contributing to metabolic block of the residual glycolytic energy production by inhibiting key enzymes (e.g., phosphofructokinase), protons also contribute to the activation of lysosomal hydrolases and lipoprotein lipase. This, coupled with proton-induced Ca^{2+} shifts and pH-induced membrane conformational changes, contributes significantly to increased membrane permeability.

As ischemia progresses, transmembrane ionic gradients are dissipated resulting in the loss of intracellular K^+ and accumulation of Na^+ and Cl^- due to the metabolic inhibition of Na^+-K^+ ATPase. It is described in detail above that this in turn leads to osmotic cell swelling. A decrease in intracellular pH also facilitates Na^+-K^+ exchange, exacerbating intracellular Na^+ overload. This in turn contributes to an increase in cytosolic-free Ca^{2+} by facilitating extracellular Ca^{2+} uptake via the Na^+-Ca^{2+} exchange mechanism. Other mechanisms, including inhibition of the energy-dependent sarcoplasm reticulum uptake of Ca^{2+} in myocardium and excitatory amino acid stimulation of neurotransmitter receptors/channels in neuronal tissue, further contribute to a massive rise in intracellular Ca^{2+}. The rise of $[Ca^{2+}]_i$ is further known to activate Ca-dependent phospholipases, phospholipase A_2, and phospholipase C, as well as proteolytic enzymes, which are also responsible for membrane injury. These enzymes do not require oxygen or energy and so may function during or after ischemia. Hydrolysis of the membrane phospholipids releases free fatty acids (FFAs), which have detergent properties that can destroy the lipid portions of all membranes. The major FFA is arachidonic acid, which can be metabolized to free radicals, prostaglandins, and leukotrienes, which can produce further changes in membrane permeability and ion distribution. It is further indicated that increasing concentrations of intracellular calcium also activate enzymes that can convert xanthine dehydrogenase to xanthine oxidase, which is known to enhance superoxide formation from hypoxanthine, especially during reperfusion.

It is now understood that calcium serves as a key messenger and modulator of intracellular signaling reactions, which affect certain enzymes, the membrane permeability, and transmitter release. Cells maintain an enormous gradient of calcium concentration across the plasma membrane (10,000:1) and the influx potential is controlled by both voltage-sensitive channels and receptor-controlled channels. Calcium efflux is via a Ca^{2+} translocase and a Na^+-Ca^{2+} exchange mechanism, both of which require energy. Also, calcium uptake into cellular organelles is energy dependent. During ischemia, Ca^{2+} enters the cell through both types of channels and, coupled with a reduced uptake by organelles, accumulates in the cytosol. This intracellular overload appears to be one of the common pathways leading to irreversible cell damage by the mechanisms summarized above.

Structural Changes

As mitochondrial oxidative phosphorylation is the first casualty of ischemia, it is not surprising that structured alterations in the mitochondrion are regarded as early and sensitive indicators of ischemia and have been the subject of extensive study. Detectable changes do not occur immediately, and first changes, manifest by disappearance of glycogen granules, lysis of cristae, and swelling, have been demonstrated after 30 to 40 minutes of ischemia in the human myocardium during cardiac surgery. At the later stages of ischemia, severe ultrastructural damage becomes evident and is characterized by extensive mitochondrial damage, pyknotic nuclei, cellular swelling, myofibrillar disruption, and the appearance of contracture bands. After 40 minutes of normothermic ischemia, irreversible changes take place and reperfusion at this stage leads to explosive cell swelling, deposition of calcium phosphate, and intense ischemic contracture resembling rigor mortis.

Vascular Injury during and Subsequent to Ischemia

In addition to the injury sustained by parenchymal cells, it is now well established that ischemic organs are subject to further modes of injury relating to vascular effects. These are collectively referred to as the "no-reflow" phenomenon and "*reperfusion injury*," which has itself been shown to be a distinct phenomenon characterized by ultrastructural, functional, and metabolic alterations.

No Reflow

A well established concern in isolated organ preservation for transplantation is that blood flow can fail to return in an organ that has suffered a period of ischemia. Clearly, this is of great importance in determining the fate of the transplanted organ, the health and viability of which depends critically upon the patency of its vascular network. Various mechanisms have been proposed to account for this phenomenon, which is also of crucial importance for the outcome of hypothermically stored organs. Contributory factors include ischemically induced vascular collapse; osmotic swelling of vascular endothelium leading to increased vascular resistance and vessel occlusion; and erythrocyte clumping producing blockage of capillaries and the formation of infarcts. The increased rigidity of red cells due to ATP depletion is considered to be a principal cause of reduced deformability and the most significant component of the no-reflow phenomenon. Weed et al. proposed that ATP normally chelates intracellular calcium and, when ATP is no longer available, calcium binds to membrane proteins, rendering the membrane more rigid. Additionally, it is known that the capillaries become increasingly leaky to protein after more than 30 minutes of ischemia, which would lead to loss of the oncotic pressure that retains fluid in the capillaries, and, hence, to an increase in the hematocrit within the vessels. This, in turn, would increase viscosity dramatically and lead to stagnation. No reflow during rewarming and reintroduction of blood into cold ischemic organs is now known to involve a network of complex interactions between vascular endothelium, blood components, and free radicals that is referred to as reperfusion injury.

Reperfusion Injury

The concept of reperfusion injury comes from the well-known fact that making a tissue hypoxic does not necessarily produce injury, but after reperfusion such tissues show marked and occasionally severe damage. Several possible interacting mechanisms of reperfusion injury are often described and include the following:

1. Cell-derived free radicals (the oxygen paradox)
2. Actions and products of inflammatory cells in the blood, especially neutrophils and platelets
3. Effects of intracellular calcium accumulation (the calcium paradox)
4. Loss of membrane phospholipids

A complete understanding of the various proposed mechanisms of reperfusion injury is far from clear and remains under intensive investigation. While a detailed discussion of the various, seemingly disjointed hypotheses is beyond the scope of this article, it is appropriate for subsequent discussion of the effects of hypothermia to include a few salient comments regarding the role of free radicals.

During the past decade, oxygen-derived free radicals (ODFR) have been the focus of attention as mediators of various tissue injuries and particularly microvascular injury. A free radical is a molecule with an odd, unpaired electron in its outer shell (denoted by a dot, thus $R^{\bullet}$), and this chemically "unsatisfied" electron renders the molecule highly unstable and reactive. Free radicals are inherently damaging since this high reactivity can precipitate chain reactions that produce increasingly reactive and toxic free radicals. Reactive species derived from oxygen are generated because oxygen normally undergoes tetravalent reduction to water by accepting four electrons simultaneously in the mitochondrial cytochrome oxidase system. However, as much as 2% of cellular oxygen undergoes univalent reduction, accepting one electron at a time and creating a superoxide anion ($O_2^{-\bullet}$, hydrogen peroxide, and eventually a hydroxyl free radical ($OH^{\bullet}$), thus the following equation:

$$O_2 \xrightarrow{e^-} O_2^{-\bullet} \xrightarrow{e^- + 2H^+} H_2O_2 \xrightarrow{e^- + 2H^+} H_2O + OH^{\bullet}$$

The hydroxyl free radical is the most reactive of all and will oxidize any organic molecule almost instantaneously. The small quantities of free oxygen radicals produced during a cell's normal metabolism are detoxified by the enzyme superoxide dismutase (SOD):

$$O_2^{-\bullet} + O_2^{-\bullet} + 2H^+ \xrightarrow{SOD} H_2O_2 + O_2$$

This protective enzyme occurs in all aerobic tissue but is found in substantial quantities only inside the cell. ODFRs are also detoxified by the naturally occurring enzymes catalase and peroxidase. During ischemia, the production of hypoxanthine is greatly increased as a result of the catabolism of ATP. In the absence of oxygen, the enzyme xanthine dehydrogenase (type XD) is converted to xanthine oxidase (type XO), which converts hypoxanthine to xanthine and this reaction also involves calcium. During reperfusion when the ischemic tissue is again exposed to oxygen, xanthine oxidase catalyzes the generation of superoxide radicals in the following reaction:

$$\text{Xanthine} + H_2O + 2O_2 \xrightarrow{XO} \text{uric acid} + 2O_2^{-\bullet} + 2H^+$$

In tissues such as the myocardium, defense mechanisms against superoxide-mediated ischemic injury are well developed in the form of scavenging enzymes. These may be antioxidants, such as glutathione peroxidase and catalase, or chain breaking antioxidants, such as SOD, ascorbate (vitamin C), and α-tocopherol (from vitamin E). The emerging role of ODFR in ischemic injury has raised the question as to whether or not injury ascribed to ischemia is in fact reperfusion injury that is initiated by ischemia, but precipitated by reperfusion. This has led to the concept of scavenging free radicals during both ischemia and reperfusion. The role of drugs like allopurinol, which inhibits xanthine-oxidase in preventing superoxide-mediated injury, is therefore readily apparent. Also, compounds that chelate those transition metals like iron — which are known to catalyze the formation of free radicals and thereby contribute to tissue injury by initiating and propagating lipid peroxidation — have been shown to ameliorate tissue damage.

For example, desferroxamine inhibits the Haber-Weiss reaction or more efficient Fenton reaction in which highly reactive hydroxyl radicals are generated when H_2O_2 accepts an electron from a reduced metal ion such as Fe^{2+}[25].

The end result of prolonged ischemia is cell death mediated by the mechanisms outlined above, which characterize the process of necrosis, or pathological cell death. An alternative mode of cell death involving a programmed or regulated process involving *de novo* gene transcription, and referred to as apoptosis, is discussed briefly below in the context of hypothermic preservation.

The culmination of the cascade of interactive ischemic events is a tissue that is unable to resume its normal function upon restoration of its blood supply. The sequence of injurious processes advances at such a rate that irreversible damage is sustained by most organs within one hour of ischemia at 37°C, and at much shorter times in highly metabolic tissues such as cardiac muscle. The role of low temperatures in protecting against ischemic injury and providing a means for preserving tissues will be considered in the remainder of this chapter.

Hypothermia in Relation to Ischemic Events

Cooling cells to varying degrees has proved to be the foundation of nearly all effective methods of viable tissue preservation. However, the consequences of cooling are not exclusively beneficial in combating the effects of ischemia, and hypothermic protection is a compromise of the benefits and detriments of cooling as outlined in the ensuing discussion.

General Suppression of Reaction Rates

The fundamental basis of all biologic and chemical processes is molecular activity and mobility, which are governed by thermal energy. This means that as temperature is lowered molecular motion is slowed. The removal of heat from a system slows down both physical and chemical processes in proportion to the loss of heat and therefore to the fall in temperature. Physical phenomena such as osmotic pressure depend solely on the rate of molecular motion so that the decrease in the rate of the process is proportional to the fractional change in absolute temperature. Many chemical reactions, however, depend upon an energy of activation, which is the minimum energy required for molecules to react. This results in a special relationship between the rate of reaction and temperature described originally by Arrhenius and outlined below. Since the processes of deterioration associated with ischemia and anoxia are mediated by chemical reactions, it has proved well founded to attempt to prevent or attenuate these changes by cooling. Biochemical processes involve molecular interactions that are invariably catalyzed by enzymes in reactions that require energy input from cellular stores such at ATP or creatine phosphate. Cooling can affect all components of these reactions including the energy status of the substrate molecules, the stability of the enzyme protein, and the capacity of the cell to supply biological energy. The rate of biophysical processes such as diffusion of ions and osmosis declines linearly with temperature by approximately 3% per 10°C. It is apparent, therefore, that biophysical events are relatively only slightly affected by the comparatively modest temperature changes imposed during hypothermic storage of tissues for transplantation. It is only at much lower temperatures that the rate of biophysical processes becomes significantly important, especially at subzero temperatures when phase changes lead to both ice formation and solute concentration changes.

By comparison, the rate of chemical reactions, including the biochemical processes that constitute metabolic activity, is slowed significantly more by a given degree of hypothermic exposure. It is well established that within the temperature range 0–42°C oxygen consumption in tissues such as kidney, liver, and heart decreases by at least 50% for each 10°C fall in temperature. Oxygen consumption (VO_2) is a reasonable measure of metabolic activity since for practical purposes, tissue and cellular stores of oxygen do not exist and cells rely upon the circulation to deliver oxygen in quantities determined by the rate of O_2 consumption. The magnitude of decrease of VO_2 by cold storage is therefore regarded as an index of the degree of reduction of metabolic activity. For tissues having a high rate of metabolism, such as the brain VO_2 at 5°C has been estimated to be 6% of the normothermic rate. The quantitative relationship between energy requirements for biochemical processes and temperature changes have been expressed mathematically in different ways:

1. The Arrhenius Relationship: Biochemical processes, in common with all chemical reactions, occur only between activated molecules the proportion of which in a given system is given by the Boltzman expression exp (-E/RT) where E is the activation energy, R is the gas constant, and T is the absolute temperature. According to the Arrhenius relationship, the logarithm of the reaction rate (k), is inversely proportional to the reciprocal of the absolute temperature:

$$-\log k = A(-E/2.3\ RT)$$

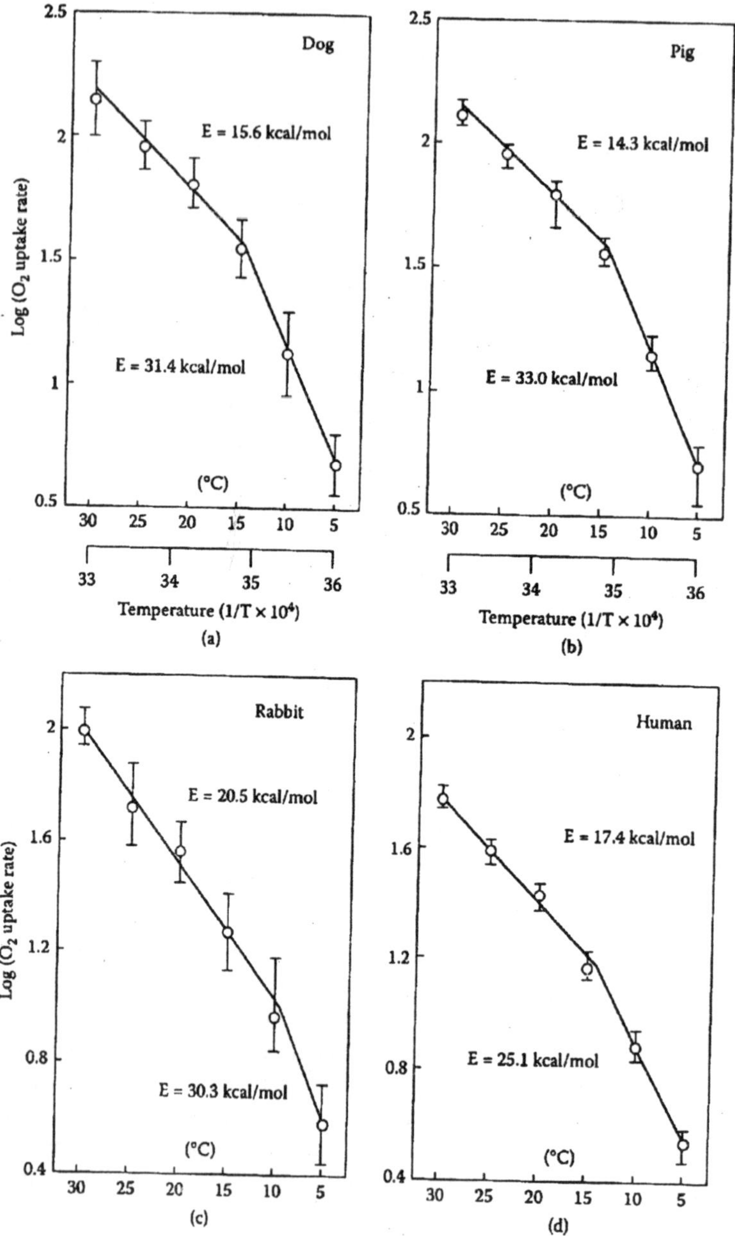

Fig. 7.5. Arrhenius plots depicting the effect of temperature on ADP-stimulated respiration in mitochondria from four different species.

A graphical plot of log k against 1/T will yield a straight line with a slope of E/2.3R if the relationship represents a single rate-limiting step. However, many examples in the literature demonstrate discrete "breaks" in the linearity of Arrhenius plots.

2. Van't Hoff Rule relates the logarithm of a chemical reaction rate directly to temperature and is commonly expressed in the form of the respiratory quotient or temperature coefficient, Q_{10}, where Q_{10} is the ratio of reaction rates at two temperatures separated by 10°C. Accordingly,

   $Q_{10} = (K_2/K_1)\ 10^{(T2-T1)}$.

For most reactions of biological interest Q_{10} has a value between 2 and 3, but some complex, energy-dependent reactions have a Q_{10} between 4 and 6, and are more likely to stop completely at low temperatures. Both Q_{10} and Arrhenius plots have been used to quantitate changes in metabolic processes occurring in biologic systems, whether they are enzyme reactions in single cells or the oxygen consumption of the entire human body. The Q_{10} for whole-body oxygen consumption is approximately 2, indicating that, in general, metabolic rate is halved for each 10°C drop in temperature. Some individual tissues, however, have been shown to exhibit a Q_{10} as high as 5 to 8, demonstrating the profound effect cooling can have on retarding reaction rates.

The impact this cooling effect has on ischemic tolerance is amply illustrated by data from hypothermic preservation of mammalian kidneys. It is explained above that kidneys can tolerate only about 45 minutes of warm ischemia before incurring irreversible injury. However, tolerance is extended to 2 hours at 15 to 25°C, 6 to 7 hours at 5 to 15°C and 12 hours at 0°C without serious injury. The same holds true for tissues that are exquisitely sensitive to ischemic injury such as the myocardium and neuronal tissue. For example, based upon the estimate that VO_2 for the brain is 6% of the normothermic rate at 5°C, Bering postulated that the brain may tolerate ischemic periods for up to 3 hours at temperatures below 5°C. This has proven to be consistent with our own recent demonstration of hypothermic protection of the heart, brain, and visceral organs during $3^1/_2$ hours of cardiac arrest and whole-body asanguineous perfusion at 7°C. Importantly, it should be pointed out here that the ischemic tolerance of organs both *in situ* and *ex vivo* is not only a function of temperature reduction *per se*, but is also maximized by manipulation of the extracellular environment of the component cells in terms of the chemical composition of the perfusate.

Metabolic uncoupling

While it is clear that cooling has a profound effect upon biochemical reaction rates and that this in turn can slow degradative processes and reduce the rate of substrate and energy depletion, it is important to realize that not all reaction rates are affected to the same degree, or even in the same manner, by cooling. For example, Southard et al. studied the comparative effect of temperature on the rate of a membrane-bound enzyme catalyzed reaction and ADP stimulation of respiration in mitochondria isolated from the kidney cortex of four species commonly used for transplantation studies, including human. Their findings which shows that the Arrhenius plots appear discontinuous with "break" points at 15°C or higher for dog, pig, and human and at 10°C for rabbit. Dog and pig mitochondria showed the greatest increase in activation energy at temperatures below the break point. In general, an Arrhenius plot with a distinct "break" has been taken to indicate that the rate-limiting step has changed but the Arrhenius Law still holds true on either side of the break. Although the interpretation of discrete changes in the slope of Arrhenius plots has been contentious, the Lyons-Raison phase change hypothesis of chilling injury has received much attention and general acclaim. This hypothesis states that at a certain critical temperature within the chilling injury range, the membrane lipids undergo a transition from a liquid-crystalline to a solid gel state. The same phenomenon has been demonstrated in the cells of mammalian organs, and the two main consequences of the transition thought to eventually result in cell injury are an increase in membrane permeability, and an increase in the activation energy of membrane-bound enzymes. While an increase in E in itself may not be damaging to a cell, it has been proposed that the damage is probably responsible for the different behavior of soluble enzyme systems and membrane-associated enzyme systems. Thus, the result would be the accumulation or depletion of metabolites at the point of entry into mitochondria. Hence, the membrane phase transitions in subcellular membranes could cause metabolic imbalance and provide one component of injury sustained by homeothermic cells during cold exposure.

One interpretation of cold resistance in various cold-adapted species is that their cell membranes maintain a greater degree of fluidity at low temperatures compared to most normothermic, non- cold adapted species. Although the role of membrane fluidity and phase transitions in cold adaptation has not been accepted unequivocally, it has been demonstrated in mitochondria from a warm-blooded species (dog) that low levels of membrane lipid perturbers such as adamantine (cyclodecane) can abolish the discontinuity in the Arrhenius plot, suggesting that membrane fluidity had been increased during cold exposure. Other lipid perturbers (e.g., Butylated hydroxytoluene) have been shown to protect homeothermic cells against cold shock, which occurs during rapid cooling from normal temperatures to 0°C. Nevertheless, this potential mode of protection for mammalian organs during cold storage has not yet been widely investigated as an interventional strategy.

The data of Southard et al., emphasize that caution should be exercised when comparisons of cellular responses to cold are made between species. It can further be appreciated that if a single biochemical process — in this case ADP-stimulated mitochondrial respiration in the kidneys from four different euthermic species — is affected differently by cooling, that it is also likely that the different biochemical reactions within a given cell will not be affected to the same extent, or in the same way, by a reduction in temperature. Figure shows schematically the large number of different, integrated chemical reactions that constitute common metabolic pathways within a cell. Each of these reactions will be affected in a different manner and to a different degree by cooling, and so the possibility exists for uncoupling reaction pathways and producing harmful consequences. This type of metabolic perturbation is largely beyond the influence of investigator intervention since it is not possible to select which processes will be depressed by cooling. It is nevertheless important for optimal preservation that any cold-induced dislocation of interconnected pathways is fully reversible or recoverable upon rewarming and reperfusion.

Optimum temperature for hypothermic storage

Cooling prolongs *in vitro* survival because it slows metabolism, reduces the demand for oxygen and other metabolites, and conserves chemical energy. However, it does not affect all reactions to the same extent, and the net result of cooling on integrated metabolizing systems is complex, not entirely predictable, and not completely understood. The application of mathematical relationships such as the Arrhenius and the Van't Hoff rules to help quantitate, predict, and understand the mechanisms of hypothermic preservation are somewhat simplistic since they relate to temperature change as the only variable. Nevertheless, this has proved useful in practice because the complexities of cooling integrated tissues and organs do not permit convenient

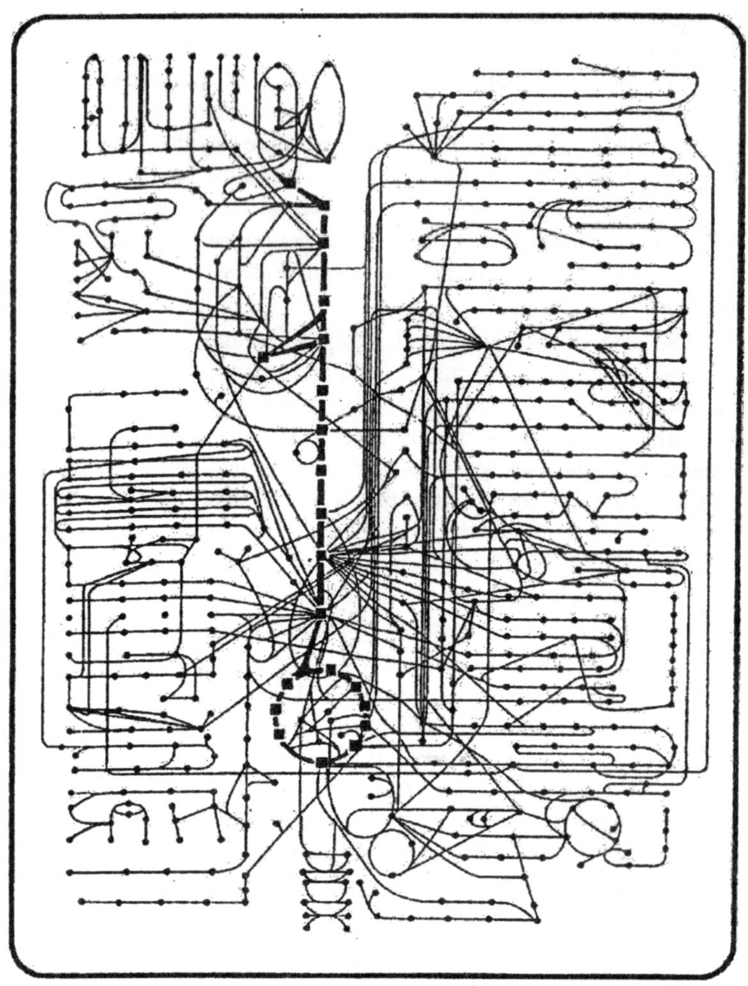

Fig. 7.6. Schematic diagram illustrating the complexity of integrated biochemical pathways in a typical cell.

separation of the interacting variables that affect the outcome of hypothermic preservation. Moreover, the effect of hypothermia *per se* on transplanted tissues and organs is confounded by the effects of prior warm ischemia and hypoxia that will undoubtedly influence the susceptibility and response of component cells to cooling.

Using tissue culture cells as an experimental model, Kruuv et al. have examined the effects of *pure hypothermia* on cell viability in the absence of any prior hypoxia. They showed that the Arrhenius plot of inactivation (killing) rates of cells exposed to reduced temperatures changes slope at approximately 7 to 8°C, implying that there are distinct mechanisms of hypothermic inactivation above and below this transition temperature. In the range of 8 to 25°C, the activation energy from the Arrhenius plot for control cells is about 15 kcal/mol, which falls within the range of temperature coefficients of metabolic processes (10–30 kcal/mol) and much lower than that for protein denaturation. Below 8°C, the magnitude of the apparent activation energy is large (-61 kcal/mol). These values have been interpreted to suggest that unbalanced metabolism is probably the rate-limiting step for hypothermic inactivation in the higher temperature range, and membrane lipid phase transition or cold denaturation of a critical protein is likely to be responsible for the strong temperature dependence in the lower range. It is apparent, therefore, that the optimum temperature for hypothermic storage will depend upon a variety of factors involving the interaction of hypothermia, the nature of the cell, and the chemical composition of its environment. This is well illustrated by reference to the voluminous literature relating to the role of hypothermia in protecting the ischemic heart and the associated design of cardioplegic solution. Although the ability of hypothermia to reduce myocardial energy demand is well established, the importance of its contribution, the optimum temperature, and the possibility that some cellular injury may be induced by the cooling have been debated and widely reported. A summary statement will suffice for the purpose of the present discussion.

The first point is that because electromechanical work accounts for approximately 85% of the oxygen demand of the myocardium, chemical arrest of the heart in diastole, with agents such as potassium or magnesium, is an important addition to hypothermia to achieve optimum preservation. Although hypothermia can abolish organized myocardial contractions, energy-consuming ventricular fibrillation can persist even at low temperatures. It is well established that irrespective of the temperature, the myocardium is best preserved during global ischemia by the combination of chemical and hypothermic arrest. With respect to temperature, the general consensus from a variety of experimental and clinical observations is that maximal myocardial preservation during ischemic arrest is best achieved in the range of 10 to 20°C. For example, Tyers et al. showed that metabolic recovery was best when the myocardium was kept at 10 to 15°C with rapid reperfusion recovery of high energy phosphates and glycogen, compared with metabolic deterioration at 4°C. Shragge et al. progressively lowered the myocardial temperature *in vitro* to 0.5°C and found no significant decrease in the concentration of ATP or the glycogen stores in the nonanoxic hearts. These findings, confirmed by others, led to the conclusion that hypothermia in the absence of ischemia is not harmful to the myocardium. Such conclusions do not, however, account for the rate of cooling that in other cellular systems is known to influence cold-induced injury by way of an ill-defined mechanism termed "*thermal shock*".

The second point to note is that it has been suggested that the safe period of ischemia can be increased by adding oxygen to the cardioplegic medium in order to satisfy the small but continued metabolic demands of the cold arrested heart. The oxygen consumption of the ischemic myocardium at 15°C is 0.27 mL O_2/min/100g tissue. Nonoxygenated crystalloid (asanguineous salt solution) cardioplegic solution administered at 10°C contains 0.86 ml O_2/100 ml of solution, and so a prohibitively large volume of cardioplegic solution would have to be injected into the coronary circulation to avoid a myocardial oxygen debt occurring within a few minutes. Moreover, it has been shown that enhanced myocardial protection can be achieved by using oxygenated media in the form of either sanguineous or asanguineous solutions. In principle, the advantage of delivering oxygen (and possibly other crucial metabolites) to the arrested heart is the maintenance of cell respiration and oxidative phosphorylation during global ischemia at temperatures that permit significant metabolism to proceed (10–20°C). This is in contrast to the strategy of tissue-suspended animation at lower temperatures (ice storage 0–4°C) where the objective is to reversibly inhibit all cellular function. The former strategy of metabolic support during hypothermic storage at ~10°C may demand continuous perfusion to supply essential substrates and remove toxic catabolic products.

Although tissue energy requirements are minimal at deep hypothermic temperatures, there are suggestions that constant supply of oxygen along with adenosine as a precursor to the ATP will result in superior ATP levels and minimum oxidative and metabolic stress in preserved tissues. However, there is no clear consensus among the research community about the need for oxygen supply during hypothermia. It has been assumed that low concentrations of molecular oxygen, such as that dissolved in organ preservation solutions, is sufficient to support the generation of free radicals during prolonged storage. Therefore, it is recognized that hypothermia may set the stage for a progressive development of tissue injury as a result of reactions and processes that occur during hypothermia, but that fuel changes that proceed for a considerable time after normal conditions of temperature and oxygen tension are resumed. Others have shown that a moderated oxygen tension is beneficial during hypothermic preservation, which suggests that oxidative stress can lead to adaptation in tissues and increased production of antioxidants. It has been shown in experimentation that rats that were gradually exposed to oxygen increased their production of antioxidants in lungs. Nevertheless, numerous investigations have suggested that oxygen supply is essential during hypothermic preservation of livers. Recent studies on survival transplantation of rat livers from donors with nonbeating hearts suggest that the saturation of UW solution with atmospheric air is a primary requirement for the preservation and restoration of ATP levels and mitochondrial functions. Previous studies by Stubenitsky et al. have also shown that the oxygenated hypothermic preservation of warm ischemic kidney slices can restore normal tissue ATP levels. More investigation is required to arrive at a consensus on optimum oxygen tension requirements that can provide superior graft function and prevent oxidative damage.

There is strong evidence, therefore, that optimum preservation of tissues and organs by using low temperatures requires careful consideration of the storage temperature in relation to other important factors including the characteristics of the cell, its environment, and the strategy adopted to effect maximum protection. The effect of hypothermic storage temperature *per se* has not been studied extensively in a wide variety of systems but the available evidence to verify this as a general effect extends beyond tissue culture cells and heart preservation outlined above. For example, in kidney preservation studies, Hardie et al. found 5°C to be superior to storage at 0.5°C and Pegg et al. showed significantly better preservation at 10°C compared with storage at 5°C. Even in whole-body protection during hypothermia, we have demonstrated improved outcome in a canine model when the nadir temperature during several hours of hypothermic cardiac arrest was 7°C compared with 1.5°C.

Effect upon Energy Metabolism

Under normal circumstances the supply of energy-rich compounds to fuel a cell's requirement for homeostatic control is continuously replenished by oxidative phosphorylation in the mitochondria. During cooling, however, there is a progressive exhaustion of chemical energy reserves in a cell despite the general suppressive effect of cooling on metabolism. Studies that clearly demonstrate the rapid depletion of adenine nucleotides during cold storage of organs at 0 to 2°C are suggestive that mitochondrial function is severely impaired by hypothermia. However, it has been demonstrated in the liver, for example, that the same tissues stored at 8 to 10°C can reestablish ATP reserves if an adequate supply of oxygen is maintained by continuous perfusion as discussed above. Moreover, it has also been established during hypothermic kidney preservation that the balance between glycolysis and complete oxidation of fatty acids at 10°C is controlled by the oxygen tension. Pegg et al. showed that glycolysis provided the principal source of energy at 10°C when the pO_2 =150 mm Hg, but that oxidation of caprylic acid provided the main fuel when pO_2 was raised to 650 mm Hg. Furthermore, it had previously been demonstrated by Huang et al. using well-oxygenated kidney cortex slices that the preferred substrate for energy metabolism was also markedly influenced by temperature. Under normothermic conditions glucose, amino acids, ketone bodies, and fatty acids were all utilized, but only short-chain fatty acids and ketone bodies were consumed at 10°C. Clearly, the effect of cooling on metabolism is complex and should not be regarded as causing a simple uniform retardation of all biochemical reactions.

The effects of cooling on mitochondrial processes are especially important for the outcome of cell preservation since it is essential for cell viability that the energy status is either maintained during storage or readily reestablished during rewarming. The crucial importance of this is readily appreciated in regard to the hypothermic storage of myocardial tissue, which is widely recognized to have special demands for its preservation compared with other organs. One fundamental difference between the heart and other transplanted organs that is reflected in the tolerance to cold ischemia is that the heart, as a contractile organ, must be able to sustain 90% of its function very soon after rewarming and reperfusion in order to be life sustaining. The heart, therefore, has much greater energy demands upon reperfusion for immediate mechanical work and adequate contractile function. The myocardium is known to be predisposed to ischemic contracture during prolonged cold storage: this is a progressive increase in myocardial stiffness with concomitant reduction of compliance and ventricular volume. The depletion of high-energy phosphate reserves causes the actin and myosin to interact, resulting in a progressive and eventually irreversible contracture of the heart. So, the basis of the problem is energy deprivation and dysregulation of intracellular homeostasis such that the onset of contracture occurs when ATP falls to less than 80% of normal values. In the early stages contracture does not necessarily imply a dead myocardium and strategies to delay the onset of ischemic contracture and promote the retention and repletion of high-energy phosphates are crucial for adequate methods of prolonged myocardial preservation.

The suppression of oxidative phosphorylation at low temperatures is indicative of a mitochondrial defect. As illustrated, oxygen consumption by isolated mitochondria decreases with falling temperature, usually with a change in the rate at about 15°C. More specifically, research has indicated that the enzymes responsible for translocating adenine nucleotides (AN translocase) across the mitochondrial membrane become ineffective at temperatures below the transition point of the Arrhenius plot. Also, the enzymes responsible for transporting NADH across the mitochondrial membrane via the malate-aspartate shuttle are ineffective at low temperatures. For example, it is known that the adenine nucleotide translocase of rat liver demonstrates an abrupt decrease in activity at 18°C, and although it is not rate-limiting at 37°C it could be limiting at low temperatures. It is now believed that the failure of aerobic metabolism during hypothermia is principally due to the inactivation of mitochondrial transport enzymes, despite the fact that it has also been demonstrated in some tissues that adenine nucleotides can be synthesized at the temperatures used for hypothermic storage, providing the appropriate substrates are present. Therefore, it is clear that the once-believed notion that it is the absolute concentrations of residual high energy phosphates that might dictate survival, and that their complete exhaustion results in loss of viability, is not tenable.

A great deal of evidence has now established the importance of the ability of hypothermically-stored cells to resynthesize energy-rich compounds during rewarming, which will be dependent upon the status of the adenine nucleotide pool remaining at the end of storage. As ATP and adenosine diphosphate (ADP) reserves are depleted during ischemia, the accumulating adenosine monophosphate (AMP) is dephosphorylated by 5'-nucleotidase enzymes to adenosine and other freely diffusable metabolites. Hence, these nucleosides are readily lost from the cell and no longer available for resynthesis of ATP. Moreover, as explained above, the synthesis of ATP depends upon active translocating processes from the cytosol into the mitochondria and vice versa, involving highly temperature-dependent enzymes. The extent of degradation of ATP (and CP) to the diffusable breakdown products is also important for another reason besides the depletion of ATP precursors that are crucial for subsequent immediate repletion of high-energy compounds. The second important aspect of this deamination pathway relates to the probable exacerbation of reperfusion injury by the generation of free radicals during and after cold storage as described above.

The complexity of low-temperature effects on mitochondrial respiration is not limited to the impairment of translocase enzymes. Recent studies have shown that other enzymes that control reactions of the tricarboxylic acid (TCA) cycle and the electron transport chain are affected differently by cold storage. In mitochondria isolated from hearts stored at 4°C for 12 and 24 hours, the rate-limiting enzyme in the TCA cycle, citrate synthetase, has been shown to be more susceptible to cold storage than the rate-limiting enzyme in the electron transport chain, cytochrome c oxidase. Such observations highlight the multifactorial nature of mitochondrial dysfunction after prolonged hypothermic storage.

It is clear from the foregoing discussion that the best methods of hypothermic preservation might depend upon the maintenance of high energy reserves, the prevention of ATP precursor depletion, and some level of continued metabolism, which, in turn, will be dependent upon the oxygen tension and storage temperature. Such considerations have led Pegg to suggest that future advances in organ preservation might be achieved by studying higher temperatures where translocase enzymes are more active and to use higher oxygen tensions. Also, the provision of purine or nucleoside precursors for adenine nucleotide repletion, and the use of pharmacological inhibitors of 5'-nucleotidase, such as allopurinol, are regarded as important components of modern- day preservation solutions and may prove to be advantageous for strategies designed to avert an energy crisis in hypothermically stored cells.

Effect upon Ion Transport and Cell Swelling

We were reminded earlier that intracellular ionic composition and volume regulation of a cell is maintained by a "pump-leak" mechanism in which membrane-bound enzymes transport various ions and solutes to counter the passive diffusion driven by chemical potential gradients. These active pumps are inhibited by hypothermia both by its direct effect on enzyme activity and by the depletion of high energy reserves as mitochondrial energy transduction fails. The effect of applied hypothermia is therefore similar to that produced by anoxia, but the mechanism is different.

This has important implications for its reversibility: even if an adequate reserve of ATP is maintained, the membrane pumps are unable to utilize ATP at low temperatures. When the temperature returns to normal the pumps are again able to use ATP and quickly restore the cell's ionic gradients. In the temperature range commonly used for preservation (0°–10°C) the activity of Na^+K^+-ATPase is essentially abolished; for example it is documented that for many cells the activity of Na^+-K^+- ATPase at 5°C is only about 1% of its normal level at physiological temperature.

The resultant passive redistribution of ions and water across the cell membrane and the concomitant change in the membrane potential has been demonstrated to be rapidly and fully reversible in the short term. The ready reversibility of all parameters upon rewarming was due to the retention of the necessary substrates and high energy compounds during hypothermic preservation so that the activity of the ATPase enzymes was quickly restored. While these radical changes in ion and water fluxes as a result of cooling are reversible in the short term, a gradual accumulation of detrimental effects eventually become irreversible in a way similar to the progression of warm ischemic events. Changes in the distributions of sodium and potassium may not cause irreversible alterations, but perturbation of normal transmembrane sodium gradients can adversely affect many secondary transport systems, such as those for glucose and amino acids as well as electrical events in excitable tissues. The various cation transport systems in cells are interrelated and energy dependent, such that all are influenced by temperature changes.

However, cation fluxes are not all affected in the same way during cooling; for example, a disparate effect of temperature on the permeability of dog erythrocytes to cations has been reported. Sodium flux was shown to increase during cooling from 37°C to 20°C and then decrease during subsequent cooling. Also, potassium flux exhibited a minimum at 12°C and then increased during further cooling, whereas water transport decreased in accordance with a typical Arrhenius relationship. This suggests that sodium, potassium, and water are transported across the red cell membrane by independent mechanisms that are affected differently by temperature changes, thus providing further illustration of the complexity of interdependence between homeostatic processes in cooled cells.

Divalent cation transport

Cells are known to have interrelated cation transport systems depending on energy supply, and are thereby affected by reduced temperatures. Moreover, it is recognized that changes in the distribution of divalent cations (Ca^{2+}, Mg^{2+}) as a consequence of ischemia, hypoxia, and even cooling are especially important in cellular injury. The belief that calcium, in particular, is a mediator of cell death is based upon accumulated evidence over several decades from observations in a wide variety of tissues and pertains to cell death from a variety of causes. During the past decade, the recognition of the central role of calcium-mediated effects in the death of cells of ischemically-sensitive tissues such as heart and brain has led to intense scrutiny of the effects of perturbations in divalent cation homeostasis. With the advent of the excitotoxic hypothesis of neuronal cell death, emphasis has shifted away from the traditional ideas of Ca-mediated injury being caused by influx of calcium into energy-compromised cells via voltage-sensitive channels, toward mechanisms involving agonist-operated calcium channels gated by excitatory amino acid receptors.

Cytosolic calcium concentration is 10,000-fold lower than extracellular concentrations, and it is now postulated that enhanced or unbalanced calcium influx across the plasma membrane represents a final common pathway in cell death mediated by various conditions that share the tendency to induce an abnormal membrane permeability for calcium. For example, excessive intracellular accumulation of calcium has been implicated as playing a pivotal role in ischemia-induced neuronal death and the evolving knowledge of the mechanisms involved include the following: alteration of electron transport in the respiratory chain, which leads to swelling and destruction of mitochondria; the release of additional excitotoxic neurotransmitters; activation of deleterious intracellular enzymes such as lipases and proteases, which break down cellular protein and lipid structures; and the formation of potentially harmful oxygen and hydroxyl free radicals and cellular depolarization.

Massive calcium accumulations also occur as a result of the so-called calcium paradox induced in cardiac cells when hearts are perfused successively with a calcium-free medium followed by a solution containing calcium. Calcium loading ensues because the calcium free perfusion dramatically increases membrane permeability to calcium. Calcium functions as a second messenger in the regulation of numerous biochemical and physiological processes, often by activating regulatory proteins such as Calmodulin, which, in turn, can activate many intracellular enzymes including protein kinases. It has been proposed that the significant protective effect of mild hypothermia against ischemic brain injury might be mediated in part by inhibition of calcium-induced effects and the prevention of inactivation of important Ca-dependent enzyme systems such as the kinases. However, it is generally recognized that cooling has no beneficial effect in preventing inhibition of the (Ca^{2+} Mg^{2+})-ATPase pump mechanism (and may even accentuate it), which is generally the most important factor in controlling intracellular Ca^{2+} concentrations. The effects of cooling on calcium homeostasis results not only from the inhibition of transmembrane pumping, but also from the inhibition of calcium sequestration by endoplasmic reticulum and mitochondria. All three pathways are involved in control of free calcium-ion concentrations, and all are sensitive to ATP depletion. Thus a reduced supply of ATP in cells stored hypothermically can result in increasing intracellular free calcium concentrations via redistribution from internal compartments, even if the external calcium concentration is lowered by perfusing with a low-calcium solution or calcium-channel-blocking drugs are added to the extracellular medium.

Calcium antagonists acting at the site of the plasma membrane will presumably be ineffective at preventing increased concentration of ionic calcium mobilized from intracellular stores. It has been discovered that the rapid influx of calcium responsible for the calcium paradox occurs via channels that are not affected by slow channel blockers, and at temperatures below 27°C changes in the cell membrane decrease slow channel currents, obviating the need for calcium channel blockers.

Proton activity changes

The scheme for the principal events of the ischemic cascade shows that elevation of the concentration of protons — i.e., increasing acidity — is regarded as a contributory central event in the process of cellular injury ensuing from O_2 deprivation and energy depletion. Moreover, reduced temperatures are also known to influence pH regulation, which is another important homeostatic mechanism for cell survival. It has frequently been reported that hydrogen ion concentration increases in a variety of mammalian cells during hypothermic storage such that tissue pH has been recorded to fall to 6.5 to 6.8 within a few hours of cold storage. Acidity is widely recognized as a hazard for cells with the accumulation of protons contributing to a variety of deleterious processes including metabolic block of glycolysis and structural damage. Destabilization of lysozomes releasing harmful proteases and catalysis of oxidative stress by mobilization of free heavy metals have been implicated as mechanisms of cellular tissue injury during acidosis.

Under physiological conditions (37°C) cells actively regulate pH within narrow limits with intracellular pH being maintained at a lower value (pH = 7.0 ± 0.3) than the extracellular fluid (pH = 7.4). The structure/function relationships of biomacromolecules, especially proteins, are governed by their tertiary and quaternary structure, which, in turn, rely upon the maintenance of charged moieties within the molecule. Pertubation of pH homeostasis in conjunction with the altered ionic environment during hypothermia can serve to markedly alter the structure/activity relationship of molecules such as enzymes that rely on electrochemical neutrality to maintain their net charged state, relative to the neutral point of water. In aqueous solutions such as the intracellular fluid, the concentrations of protons (H^+) and hydroxyl ions (OH^-) are determined by the ionization constant of water (pKw), which increases as temperature decreases. Intracellular electrochemical neutrality (equal concentrates of H^+ and OH^-) is therefore maintained at reduced temperatures only if pH rises in concert with pKw to

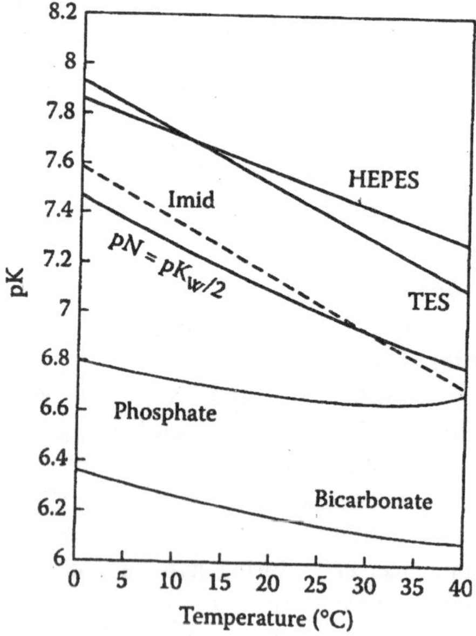

Fig. 7.7. The relationships between pH and pK as a function of temperature relative to acid-base control in biological systems during hypothermia.

maintain a constant OH^-/H^+ ratio, i.e., the quantity of protons needed for neutrality falls as temperature decreases. It is important to understand that a given pH value is not a measure of electrochemical neutrality unless it is related to a specific temperature, because neutrality is also dependent upon the concentration of hydroxyl ions. Hence it is only at 37°C that an extracellular pH of 7.4 yields a "neutral" intracellular pH that is optimal for physiological function. In other words, if pH does not rise as temperature is lowered there will be a relative excess of protons despite an apparently "normal" pH value. The disturbance of intracellular neutrality by the accumulation of H^+ during hypothermic ischemia not only has a profound effect upon macromolecular structure and poisons the active sites of enzymes, but also causes metabolites to lose their charged state, and therefore are able to diffuse down concentration gradients as nonionized lipophilic molecules. The depletion of important metabolites for the regeneration of high-energy phosphates during reperfusion is therefore exacerbated by relative acidity. The Donnan equilibrium responsible for maintenance of transcellular ion gradients and cellular water content is also dependent upon electrochemical neutrality within the cell.

At the systemic level, a failure to understand that it is the OH^-/H^+ ratio that is the critical determinant of protein structure and enzyme function, rather than the pH value *per se*, has been responsible for much misunderstanding and controversy about acid-base management during hypothermia.

Acid-base Regulation during Hypothermia

It has been established in ectothermic (cold-blooded animals), and in the blood of warm-blooded animals cooled in a closed system that does not permit gas exchange, that pH rises in parallel with the neutral point of water (pN) during cooling in the range 0°C to 40°C (the Rosenthal relationship). The rate of change of pH with temperature is -0.0157 pH units/C° and is referred to as alpha-stat pH regulation in recognition of the fact that both intracellular pH and blood pH buffering is dominated by the degree of ionization (α) of the imidazole moieties of proteins.

On the basis of both *in vitro* and *in vivo* experiments it has been generally accepted that acid-base regulations in nearly all vertebrates are consistent with primary regulation of α-imidazole resulting in a stable OH^-/H^+ ratio and the observed change in blood pH with temperature. Intracellular pH is close to neutrality and closely parallels the rise in pN, whereas in ectotherms and mammalian blood *in vivo*, body fluid pH is maintained at higher levels, i.e., it is more alkaline, than intracellular pH. The purpose of maintaining an alkalotic extracellular milieu may be to provide the cell with a proton-sink for the acidic products of its metabolism. Although this pH management strategy is the most prevalent in the animal kingdom, an alternative process has evolved in hibernating mammals: their metabolism and metabolic function continue at a body temperature as low as 5°C and they maintain arterial pH at 7.4, irrespective of systemic temperature (pH stat regulation).

A detailed discussion of these pH-regulatory phenomena is beyond the scope of this chapter, but the implications of the choice of strategy that will deliver optimum protection during clinical hypothermia — or hypothermic preservation of isolated organs for transplantation — has been addressed in numerous publications over the past few decades. Although these considerations have sometimes led to a controversial debate, there are persuasive arguments with supporting evidence in favor of the alpha-stat strategy. Again, examples from studies on exquisitely sensitive ischemic tissues such as heart and brain are of particular interest. It has been shown in numerous studies that the electrical stability, contractility, and hemodynamics of the heart is better preserved during hypothermia when the a-stat scheme is adopted, as opposed to constraining pH to 7.4. Concerns for brain protection during hypothermic cardiopulmonary bypass have led to studies of the effect of pH and temperature on cerebral metabolism and cerebral blood flow with the latter being appropriately maintained during a-stat management. In a direct comparison of the α-stat versus pH-stat management of dogs subjected to 60 minutes of cold (17°C) ischemic circulatory arrest, it was demonstrated that the α-stat strategy resulted in better protection of ischemic tissues than in those animals whose pH was maintained at 7.4 throughout. Improved protection was manifest as a better cardiac output, twice the cerebral blood flow, lower peripheral resistance, and a significantly better postischemic ventricular performance on rewarming.

Strategies of pH management for optimum hypothermic preservation of isolated organs have not yet been studied extensively, but if α-stat regulation proves to be generally beneficial to maintain a constant degree of alkalinity between the extracellular and intracellular compartments to preserve electrochemical neutrality during cooling, the question of appropriate pH buffers needs to be carefully considered. The intrinsic physiological buffers phosphate and bicarbonate do not retain their relative buffering capacity during cooling, due to the temperature coefficients of their dissociation constants (dpK/dt). In sharp contrast, the imidazole group of the amino acid histidine retains effective buffer capacity over the entire temperature range because its dpK/dt is closely aligned with that of water. In synthetic systems such as manmade organ preservation media, or hypothermic blood substitutes, other buffers are available with equivalent or greater temperature coefficients to that of the pK_w for water. For example, the aminosulfonic acid buffers introduced by Good et al. have been shown to possess superior buffer capacity and temperature coefficients for applications involving hypothermia. For example, the zwitterionic aminosulfonic acid compound HEPES* has found widespread *in vitro* use as a biocompatible buffer and the close match of its pKa/dt with that of water and imidazole.

Moreover, the aminosulphonic acid buffers may be a better choice than imidazole/histidine for preservation solutions since imidazole has been criticized as being too reactive and unstable to be a satisfactory biological buffer and has been shown to uncouple electron transport from phosphorylation. Moreover, we have recently shown unequivocally that the ampholyte HEPES provides superior buffering capacity and efficiency compared with histidine for pH regulation in the critical range of pH 7.0–7.8. The efficacy of improved buffering capacity using aminosulfonic acid buffers such as Bicine and Tricine as additives to UW organ preservation solution has also been demonstrated to improve the metabolic status of hypothermically stored livers. In this study Bicine and Tricine provided a higher buffer capacity and greater protection than histidine. HEPES has been documented to be highly effective in combating the alterations in acid-base homeostasis of ischemic hearts. We have demonstrated the effectiveness of a variety of these sulfonic acid buffers in cryobiological applications and consideration of their use *in vivo* has recently been advocated

Control of acidosis in the cells of preserved organs will be impacted by the exchange of protons and buffer species between the intracellular and extracellular compartments. With respect to external buffering power and intracellular pH, Garlick et al. addressed the question of whether protons produced within ischemic cells are transported to the extracellular space. Such an export process would slow down as extracellular pH decreased. They hypothesized that if the external pH was maintained by increasing external buffering, the proton export could continue longer, thereby reducing the fall in pH_i.

They provided some support for this hypothesis using hearts perfused with Krebs-Henseleit supplemented with HEPES. Clearly, intracellular pH (pH_i) can be influenced by external buffers if there is exchange of intracellular buffer species such as phosphate and/or protons. Moreover, there is the possibility that external buffers can permeate into the intracellular space and thereby directly act as pH_i buffers. As far as we are aware, there have been very few specific studies to examine the permeation of buffer species into cells. However, relatively small organic compounds such as histidine (155 daltons) and the amino-sulphonic acid buffers (approximately 160 to 230 daltons; e.g., BICINE = 163 daltons and HEPES = 238 daltons) might be expected to permeate during prolonged hypothermic exposure. There is recent evidence that much larger molecules such as the disaccharide trehalose (342 daltons) can permeate into cells as a result of hypothermia-induced phase changes in the plasma membrane.

Many of the reported studies purporting to examine changes in pH_i have relied upon NMR, but Lareau et al. have cautioned about the interpretation of intracellular pH changes based upon ^{32}P NMR spectroscopy because inorganic phosphate leaks from cells during prolonged hypothermic storage. For this reason, it is not always possible to follow the time course of pH_i from the chemical shift of the Pi peak.

Effect of Hypothermia on the Generation of Free Radicals

The emerging role of oxygen derived free radicals (ODFR) in tissue injury and its participation in reperfusion injury is mentioned above. Important questions that arise in the context of cold storage of tissues and organs include whether free-radical–mediated tissue injury proceeds during cold ischemia, and what effect temperature reduction has on the processes of free radical generation and the mechanisms of tissue injury. Fuller and Green and their colleagues have considered these questions; their work provides a basis for the summarizing statements recorded here.

In essence, it is recognized that cooling increases the susceptibility of cells to produce free radicals and attenuates the natural defense mechanisms by which cells normally deal with the low- level free radical production in metabolism. Due to a lower activation energy, free radical reactions are depressed less by temperature reduction than by the enzymatic processes used to scavenge them. As outlined above, the highly reactive radicals can react in a nonenzymatic way with a variety of biomolecules, including membrane lipids, proteins, and nucleic acids in order to become energetically more stable species and, in the process, generate other radicals at the site of attack. A chain of such reactions can be established, especially when catalysts are present in the form of species capable of redox cycling, i.e., shuttling between oxidation states by accepting and donating electrons. Transition metals such as copper and iron — which are fundamentally important in cell metabolism as intrinsic components of enzyme systems like the cytochromes of the mitochondrial electron transport chain — are effective catalysts of such radical chain reactions.

Loss of homeostatic control of transition metals during cold ischemia is therefore a potential hazard to cells because of the promotion of free radical production and subsequent tissue injury during either cold storage or reperfusion. It is the balance between production and removal of free radicals that is crucial to the cell such that a combination of excessive radical generation and hindrance of normal defense mechanisms can redirect the processes in favor of injury. There is evidence that some of the natural defense mechanisms against free radicals become depleted during cold storage, and that the addition of natural pharmacological scavengers including SOD, catalase, or mannitol to cold storage media may improve the viability of hypothermically stored organs.

Although it is possible for the production of injurious free radicals to be enhanced during cold storage, it is important to appreciate that the resultant cell damage may not occur entirely at the low temperature. On the contrary, there is a growing body of evidence that reintroduction of oxygenation via a regular blood supply upon rewarming and reperfusion provides a powerful impetus for further oxidative stress. A principal pathway is the stimulation of enzymically driven radical reactions such as the xanthine/xanthine oxidase system involving the interaction of ATP catabolic products with molecular oxygen as discussed in an earlier section.

Vascular endothelial cells are thought to be particularly vulnerable to the type of injury mediated by free radical generation due to this so called "*respiratory burst*" mechanism. Nevertheless, it is known that low concentrations of molecular oxygen such as what is dissolved in organ preservation solutions is sufficient to support the generation of free radicals during prolonged storage. Therefore, it is recognized that cold exposure may set the stage for a progressive development of tissue injury as a result of reactions and processes that occur during hypothermia, but which fuel changes that proceed for considerable time after normal conditions of temperature and oxygen tension are resumed.

Unrestricted bursts of free radical activity are known to cause membrane damage to a variety of cellular components due to the hazards of lipid peroxidation. While cells employ a number of repair mechanisms to recover from this type of injury, cell survival depends upon whether slavage pathways are overwhelmed or whether a point of irreversible damage is reached during the storage/reactivation process so that cell death becomes inevitable. During the transition from reversible cell injury to irreversible cell death, calcium is strongly implicated as an important link between free-radical-mediated changes and eventual cell necrosis. It is clear from the foregoing discussions that in tissues exposed to cold there is potential for unbalanced free radical reactivity and elevated intracellular free calcium levels, and subsequent warming and reperfusion is likely to greatly potentiate these adverse events.

Evidence linking cold ischemic damage and oxidative stress has come from studies involving a variety of tissues, most notably the heart, kidney, and liver. In recent years, strategies aimed at circumventing oxidative stress and reperfusion injury under both hypothermic and normothermic conditions have understandably been focused upon the roles of antioxidants and calcium channel blockers. An example in which this approach has been adopted is the development of the "Carolina Rinse Solution" designed specifically to improve the preservation of cold-stored livers by protecting against reperfusion injury.

Other studies have indicated that antioxidant supplementation of existing cold storage media such as UW solution may be advantageous in preserving organs with increased oxidation stresses sustained from suboptimal donors in clinical practice. Nevertheless, a clear picture for unequivocal benefits of using agents such as antioxidants and calcium channel blockers in a clinical setting has yet to be established. Some insights into the cellular and molecular mechanisms of cold storage injury with an emphasis on oxidative stress have recently been reviewed by Rauen and DeGroot.

Structural Changes

The interrelationship between structure and function is fundamentally important for cellular homeostasis, which is governed by an intricate array of biochemical, physiological, and biophysical processes. Moreover, these processes are compartmentalized within the cell such that the structure of biological membranes and the cytoskeleton are integral components of cell viability and vitality.

The sensitivity of biological structures to temperature change is well known and the thermal denaturation of proteins in particular is well documented. Thus, most proteins, as well as nucleic acids and many polysaccharides, are able to exist in their biologically active states only within a limited temperature range that is characteristic of the macromolecule and its environmental conditions such as ionic strength and pH. While a great deal more is known about heat denaturation of proteins at elevated temperatures, there are well described examples of cold denaturation involving the spontaneous unfolding of proteins or dissociation of the multi-subunit structure into biologically inactive species, which may or may not reassemble on rewarming to normal temperatures.

It was mentioned earlier in the discussion of metabolic uncoupling that membrane components are also affected by cooling, such that membrane properties relating to both the selective diffusion barrier to solutes and active regulation involving membrane-associated proteins (e.g., ion transporters) are altered by cooling. The effect of cooling on the thermophysical properties of membrane lipids is generally regarded as the fundamental basis of temperature effects upon membrane structure-function relationships.

In essence, it is well established that it is the fluid mobile lipid phase of membranes that supports their functional orientation and conformational movements, including that of integral or transmembrane proteins. During cooling, individual phospholipids undergo an abrupt change from a disordered fluid, or liquid crystalline state, to a highly ordered hexagonal lattice (gel state) at a specific transition temperature. Biological membranes comprise complex mixtures of lipids such that sharp transitions between the liquid crystalline and gel phase do not occur, but rather the membrane exhibits a phase transition temperature range. Moreover, compositional characteristics such as the cholesterol: phospholipid ratio in the membrane can serve to broaden or "smooth out" the phase transitions.

In a past review of the effects of cooling on mammalian cells, Fuller provided a concise yet informative synopsis of these membrane-orientated effects of reduced temperatures. It has been postulated on the basis of measurements in both model systems and intact cells that a phase separation occurs within the plane of the membrane as cooling proceeds. Such cold-induced changes in the degree of membrane fluidity render it thermodynamically unfavorable for membrane proteins to remain in the gel phase, such that they may be redistributed laterally into regions of low-meltingpoint lipids that remain in the liquid crystalline phase. This process is thought to result in packing faults due to the development of lipid-rich and protein-rich microdomains in a membrane undergoing phase transitions.

One consequence of this could be a change in membrane permeability and alteration in the solute barrier function of the membrane. Phase separation of mammalian membranes have been demonstrated at temperatures in the region of 10°C and below, and this correlates with the trend towards increased permeabilities to both ions and even large molecules such as proteins during cooling. In addition to phase separations another form of cold-induced membrane damage includes the actual loss of membrane phospholipids, which is intuitively more deleterious than phase changes that may largely be reversible.

Thermal shock

It has been discovered in some cells that the rate of cooling is also a determinant of injury. The phenomenon, which is not well understood, is referred to either as thermal shock, cold shock, or chilling injury. The phenomenon is distinct from the well-characterized effects of cooling rate in frozen cells. The pure effects of cooling, in the absence of solute concentration changes or ice formation, are thought to be related to the thermotropic properties of cell membranes involving phase transitions as discussed earlier. It has been proposed that rapid cooling in the absence of freezing can cause mechanical stresses on membranes induced by differential thermal contraction.

Stress proteins

The change in the structure of a protein is a common response to stress. Cells accumulate incorrectly folded proteins as a consequence of stresses such as hypoxia or temperature change. Proteins that experience unfolding or conformational changes do not remain soluble and are transformed into denatured proteins. It is now well established that one of the defense mechanisms adopted by cells to counteract such effects of stress is to synthesize new proteins commonly referred to as "*stress proteins*" or "*heat shock proteins*" (HSP) because of their increased synthesis by many cell types after exposure to elevated temperatures. The general role of this group of highly conserved stress proteins is believed to be homeostatic in that they protect the cell against the harmful consequences of traumas and help promote a quick return to normal cellular activities once the stress has terminated. The principal mammalian HSPs are classified into families on the basis of their molecular weights and sequence homologies, and include the 8 kD protein ubiquitin, small HSPs (20–28KD); HSP 60; HSP 70; and HSP 90. The HSP 70 family is the most widely studied. These proteins appear to bind to denatured and unfolded proteins and prevent further aggregation and precipitation. Some HSP 70 proteins are always present in cells and have been assigned a role as "*molecular chaperones*" because they mediate the correct assembly and folding of proteins. The general protective function of "stress" proteins is also thought to be mediated by their chaperone-like ability to associate with other proteins in the cell and modify their function and fate.

The induction of stress proteins in response to cold (cold shock proteins) has been identified in invertebrate species such as bacteria and yeast. Until recently it was unknown whether similar families of cold shock proteins are induced in mammalian cells. However, the contention by some that "recovery of cells from cold temperatures does not result in synthesis of specific shock proteins" will clearly need to be reevaluated in the light of recent evidence demonstrating that cold shock induces the synthesis of stress proteins in a variety of mammalian cells and organs. Moreover, the natural response to heat stress by synthesis of "*shock proteins*," a phenomenon referred to as thermotolerance, has been explored as a potential interventional strategy to protect against induced ischemia and reperfusion injury.

The idea is to precondition cells to better withstand hypothermia by previous exposure to a less severe stress, or sublethal dose of the same stress. It has been demonstrated in both cells and organ systems that "*stress conditioning*" can induce cold tolerance in mammalian systems. Cytoprotection is achieved by way of a complex adaptation in cellular metabolism analogous to that seen in natural thermotolerance. A dominant metabolic change associated with hyperthermia-induced cytoprotection is the increased expression of heat-shock genes. Recent studies have confirmed in diverse animal models that the levels of heat shock proteins are increased manyfold in heat-shocked animals, and that the stress conditioning led to significantly improved functional recovery and decreased cell death in various tissues after subsequent cold ischemia. It is particularly noteworthy that in each of these mammalian models a positive temporal association was demonstrated between the functional protection of stress conditioning and enhanced expression of inducible heat-shock proteins.

Cytoskeleton

Cold induced changes of cellular structural proteins that constitute cytoskeletal components such as microtubules have been known since the mid 1970s. This cold sensitivity appears to be mediated by depolymerization of component polypeptide units and, in general, is readily reversible upon rewarming. Nevertheless, the possibility exists for reassembly of the microtubules in a way that results in cellular abnormalities.

Apoptosis vs. necrosis in cold-induced cell death

Injured or dying cells exhibit characteristic changes in cell morphology and, as described above, the culmination of the deleterious events comprising the ischemic cascade is characterized by marked structural changes in cells and eventual cell death. It is now known that there are two distinct ways in which cells may die: necrosis — caused by a general failure of cellular homeostatic regulation following injury induced by a variety of deleterious stimuli including hypoxia, toxins, radiation, and temperature changes — and apoptosis, or "*programmed cell death*," which is a regulated process distinguishable from necrosis by numerous morphological biochemical and physiological features. In regard to structural changes, apoptosis involves distinct morphologic features including blebbing of the plasma membrane (but not loss of integrity), cell shrinkage with the formation of apoptotic bodies, chromatin condensation, and internucleosomal DNA fragmentation in a specific nonrandom pattern. These characteristics are distinguishable from necrotic changes that involve degenerative processes such as cell and organelle swelling, rupture of plasma and lysosomal membranes, and a random pattern of DNA degradation.

While many of the diverse stresses known to cause necrotic cell death have also been reported to induce apoptosis in a variety of cells, the role of low temperatures as a possible stimulus of programmed cell death has only recently begun to emerge. The earliest reports appeared in 1990 when it was shown that cultured mammalian fibroblast cells at the transition from logarithmic to stationary growth were killed by brief exposures to 0°C and rewarming at 37°C. The observed cell killing required only a few minutes of hypothermic exposure and the affected cells exhibited characteristics of apoptosis. Moreover, the kinetics of cell death served to distinguish cold-induced apoptosis from the lethal effects of longer-term cold storage (hours or day), and also showed characteristics inconsistent with direct chilling injury, or cold shock. Most recently, we have reported that apoptosis can be detected

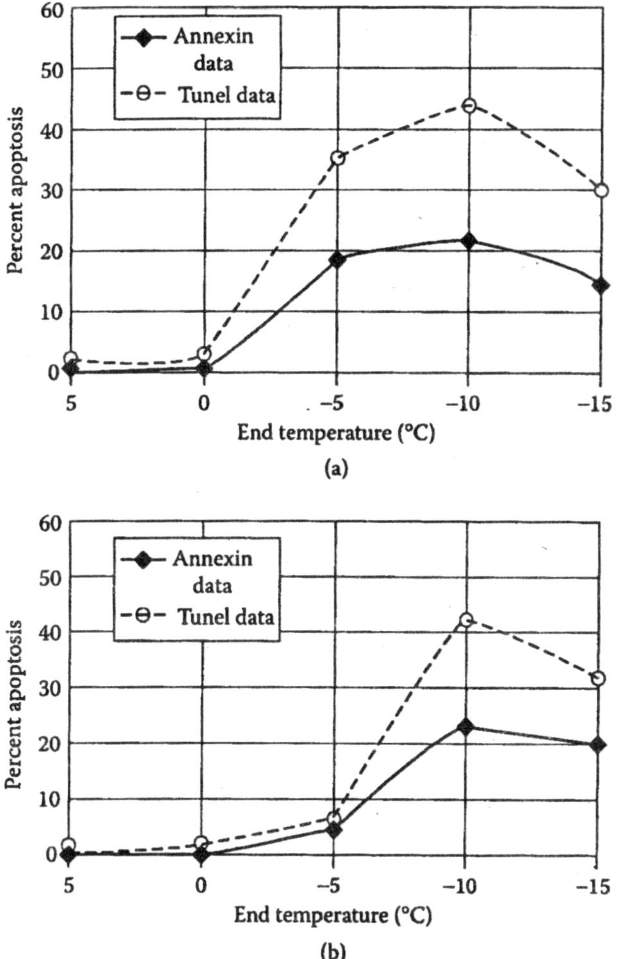

Fig. 7.8. Extent of apoptosis detected in human coronary artery endothelial cells (a) and arterial smooth muscle cells (b) as a function of exposure temperature.

in a significant proportion of both vascular smooth muscle and endothelial cells subjected to brief (30 to 120 seconds) cooling. Both cell types, apoptosis was not found after exposure to hypothermia in the range +5 to 0°C, but was induced at lower temperatures with a maximum response occurring at -10°C.

There is increasing evidence, therefore, that apoptosis may represent another manifestation of cold injury. However, it is clear that all cells cannot be as susceptible as those used in Nagle's study or hypothermic preservation for practical lengths of time would not be possible, which is not the case. Susceptibility to cold-induced apoptosis is likely to depend critically upon the cell growth cycle as demonstrated in Nagle's study, and the temperature of exposure as demonstrated by Tatsutani's study. Recent studies in cultured hepatocytes and liver endothelial cells have indicated that reactive oxygen species might play a key role in mediating cold-induced apoptosis.

It is becoming increasingly more evident that apoptosis plays an integral role in cell death induced by the rigors of both hypothermia and cryopreservation. More specifically, apoptosis has been identified to be directly associated with delayed-onset cell death (DOCD). This is defined as death associated with cold exposure that is not apparent immediately upon rewarming, but extending over the postexposure recovery period. Recent research into the causative apoptotic and necrotic pathways responsible for low-temperature-induced DOCD has identified the contribution of multiple apoptotic pathways, including receptor- and mitochondrial-induced apoptosis. Investigations into these pathways, their progression, and their induction stressors has begun to facilitate new methods for improving preservation efficacy through the modulation of the cellular and molecular responses of a cell undergoing preservation (both hypothermic and cryopreservation).

Incorporation of specific apoptotic protease inhibitors in preservation media has now been reported to markedly improve the survival of a variety of cells and tissues. Furthermore, investigation into the modification of the carrier medium from that of standard extracellular-type culture media — with or without cryoprotectants — to that of specifically designed intracellular-type preservation solutions, such as Hypothermosol and Unisol, have led to studies showing significant improvement in preservation efficacy. In a series of independent studies based upon the HypoThermosol formulation, the hypothesis that cell preservation in "Intracellular-type" solution experience reduced hypothermic stress during cryopreservation was tested and yielded higher cell survivals. Moreover, the new generation of solutions has been shown to provide better long-term hypothermic preservation compared to the industry standard (UW-ViaSpan). For example, human liver cells after just three days storage in UW at 4°C die within three days of return to physiological conditions. In marked contrast, cells stored in HypoThermosol solutions survive and propagate from three to five days later. We have demonstrated similar comparisons with a variety of other tissues and cell types stored in Unisol at 4°C. Moreover, when experimental inhibitors of apoptosis were added to HypoThermosol, human liver cell lines survived up to eight days at 4°C, compared to only 24 hours for cells stored in ViaSpan.

These and several ensuing studies have provided strong evidence in support of this concept, which has now opened the door to a new approach to modulating preservation injury by focusing on molecular mechanisms associated with cold-induced cell death. Even before this latest flurry of studies that elucidate the molecular mechanisms of cellular demise — and thereby identify ways of circumventing the problems — some of the principles were embodied in the rational design of the baseline media. More specifically, the designs of hypothermic blood substitutes such as Hypo-Thermosol, and the newer formulation Unisol, both incorporate some components that possess recognized antioxidant activities and hence implied anti-apoptotic activity. For example, reduced glutathione is a component of both formulations as a multifaceted molecule that is also known to fulfill a natural role in the regulation of apoptosis.

Since apoptosis is proving to be a distinct and significant factor in limiting hypothermic storage of cells and organs, strategies aimed at preventing apoptotic cell death should be pursued. Adding anti-apoptotic compounds, such as LXR-015, to hypothermic storage media has been reported to improve hypothermic preservation of rat liver in EuroCollins solution. A variety of low-molecular-weight compounds have now been identified and used to either enhance or inhibit programmed cell death such that apoptosis is now amenable to pharmacological intervention.

Once a cell has been exposed to a death stimulus one of two main pathways becomes activated: the intrinsic mitochondrial pathway or the extrinsic death receptor pathway. The progression of apoptosis can be halted at numerous checkpoints along the apoptotic pathways. Two key areas of research that have proven successful in extending the life of cells in culture involve the inhibition of mitochondrial dysfunction and the inhibition of caspases. Apoptosis may be inhibited through the genetic engineering of mammalian cells to over-express an anti-apoptotic gene (such as the Bcl-2 gene), or through the addition of a chemical inhibitor to the environmental medium — whether it be a culture medium or a preservation solution. As an alternative to pharmacological intervention, it has been suggested that gene therapy could provide the means to modify grafts such that they become less susceptible to preservation injury associated with apoptosis. This novel approach involves transfecting murine livers with a recombinant adenovirus vector encoding the Bcl-2 gene to reduce apoptosis during the preservation time (the Bcl-2 gene product has been shown to promote cell survival by inhibiting the process of apoptosis). Grafts expressing Bcl-2 showed significant reduction of enzymes associated with liver damage compared with control grafts. However, the genetic engineering approach is less amenable to practical intervention for the purpose of improving biopreservation.

Interventional Control of the Extracellular Environment to Optimize Preservation

This section outlines the cellular effects of hypothermia in relation to an ischemic or hypoxic insult. Hypothermia is the bedrock of all useful methods of preservation and this has proven to be most effectively applied by controlling the extracellular environment of cells directly — and the intracellular environment indirectly — during cold exposure. It is beyond the scope of this chapter to deal with the strategies that have been adopted to maximize the benefits of cooling for tissue and organ preservation, but some salient points will be offered as a prelude to discussion in ensuring chapters. Moreover, other authors have reviewed in detail the approaches that have been taken to design interventional techniques for hypothermic preservation.

It is mentioned above that control of the extracellular environment of cells to optimize preservation may be based on one of two strategies: static cold storage, or flush preservation, and low-temperature continuous perfusion. A third technique, involving gaseous perfusion (retrograde oxygen persufflation), also exists, but it remains largely experimental.

Flush Cold Storage

It will be apparent from the foregoing discussion that these different strategies call for different approaches to interventional control of the extracellular environment in order to optimize preservation. In principle, flush preservation is based on the premise that reducing temperature so that it is near — but not below — the ice point (0°C) precludes the need to support metabolism to any significant extent, and that the correct distribution of water and ions between the intracellular and extracellular compartments can be maintained by physical rather than metabolic means. Since during the period that metabolic pumps are inactivated the driving force for transmembrane ion flux is the difference in ionic balance between intracellular and extracellular fluid, and driving force for water uptake (cell swelling) is the impermeant intracellular anions, these changes can be prevented or restricted by manipulating the extracellular environment to abolish the chemical potential gradients. On this basis a variety of flush, or organ wash-out solutions, have been devised and evaluated for cold storage; these are often referred to as "*intracellular-type*" solutions due to their resemblance in some respects to intracellular fluid. The principal design elements of the intracellular-type flush solutions have been to adjust the ionic balance (notably of the monovalent cations) and to raise the osmolality by including an impermeant solute to balance the intracellular osmotic pressure responsible for water uptake. However, the most important factor for the efficacy of cold flush solutions appears to be the prevention of cellular edema by inclusion of impermeant solutes since it has been established that ionic imbalances, especially potassium depletion, are readily and rapidly reversible.

Attention to the biophysical properties of intracellular-type flush solutions to restrict passive diffusional processes has unquestionably led to the development of techniques during the past few decades that have provided the basis of clinical organ preservation. Nevertheless, it is recognized that further optimization of cold flush solutions can be achieved by inclusion of biochemical and pharmacological components that will be effective in counteracting the deleterious effects of ischemia and reperfusion injury. To some extent, this approach has been incorporated in the design of the University of Wisconsin organ preservation solution which has become the most widely used solution for cold flush preservation of kidneys, livers, and pancreata.

With due consideration for the effects of ischemia, hypoxia, hypothermia, and reperfusion injury on cells, coupled with the proven efficacy of various existing organ preservation solutions, a general consensus of the most important characteristics in the design of hypothermic storage solutions has emerged.

Table 2.1. Desirable properties of a hypothermic preservation solution or blood substitute

1. Minimizes hypothermically-induced cell swelling
2. Prevents expansion of the interstitial space (especially important during perfusion)
3. Restricts ionic imbalances
4. Prevents intracellular acidosis
5. Prevents injury from free radicals
6. Provides substrates for regeneration of high energy phosphate compounds during reperfusion

Continuous Hypothermia Perfusion Preservation

The desirable properties of hypothermic solutions applicable to controlling the extracellular environment by way of continuous perfusion techniques. In contrast to static cold storage, continuous perfusion is usually controlled at around 10°C and is based upon a different principle: it is generally assumed that a moderate degree of cooling will reduce metabolic needs but that continuous perfusion is required to support the suppressed metabolism and remove catabolic products. Because it is assumed that sufficient metabolic activity remains to actively regulate a near-normal cell volume and ionic gradients, the perfusates are generally acellular, isotonic, well-oxygenated solutions, having a composition that more closely resembles plasma than intracellular fluid. Such perfusates are therefore designated as "*extracellular-type*" solutions. They are perfused through the vascular bed of an organ at a pressure sufficient (typically 40 to 60 mm Hg) to achieve uniform tissue distribution. To balance this applied hydrostatic pressure and prevent interstitial edema, oncotic agents such as albumin or synthetic macromolecular colloids are incorporated into the perusates. Substrate support of the remaining metabolism at ~10°C is also an important consideration and it has been shown in several organs that high energy adenine nucleotides can be synthesized during either hypothermic perfusion preservation, or oxygenated hypothermic reperfusion as discussed in an earlier section above.

In addition to the principal objective of supporting metabolism, continuous perfusion also provides other advantages over flush preservation. These include the washing out of accumulated lactate and protons, which removes the metabolic block on glycolysis. This is thought to be especially beneficial for organs that have suffered prior warm ischemia. Perfusion also facilitates the removal of erythrocytes from the microcirculation and helps to maintain vascular patency during prolonged storage. Continuous perfusion has been shown to provide the best means of achieving prolonged hypothermic preservation (e.g., 3 to 7 days for kidneys), but concerns for damage to the vascular endothelium during prolonged perfusion may be a limiting factor. Despite the apparent advantages of continuous perfusion over the flush techniques, static cold storage remains the method of choice at most clinical transplant centers. This is largely because of its simplicity and convenience over the complexity, expense, and risks of vascular damage associated with prolonged continuous perfusion technology. Nevertheless, in the context of utilizing expanded criteria donor organs to make available more organs for transplantation, machine perfusion technology is considered essential and there is currently a resurgence of clinical interest in these techniques.

Approaches toward Universal Tissue Preservation

Although it has been experimentally verified that cell metabolism continues at temperatures as low as 10°C and that adenine nucleotides can be resynthesized during hypothermic preservation if appropriate substrates are provided, it is considered unlikely that this level of metabolism can prevent transmembrane ion and water movements: this is due principally to the temperature sensitivity of the active pumps. Hence, some advocates of continuous perfusion have modified the perfusate accordingly by increasing both the K^+ concentration and the osmolality Similarly, modification of cold flush solutions can be considered to circumvent some of the identified limitations of that approach. For example, it is described earlier that the lack of support of metabolism during ice-storage can be addressed by raising the temperature of storage, and by incorporating biochemical substrates and raising the oxygen tension to promote adenine nucleotide repletion. Also, the use of pharmacological agents such as inhibitors of 5'-nucleotidase (e.g., allopurinol) has been advocated as a means of averting adenine nucleotide depletion.

Therefore it is apparent that optimum control of the intracellular and extracellular environment of cells during hypothermia depends upon the interaction of a variety of factors that include temperature, oxygen tension, acidity, osmotic pressure, and chemical composition of the perfusion fluid or wash-out solution.

As transplantation science progresses and the demand for effective methods of preservation of an increasing variety of tissues and organs escalates, so the search for a "universal" scheme of cold storage has become an important goal. The modern history of organ preservation since the early work of Belzer's group and Collins's group in the 1960s has shown that the original solutions devised for kidney preservation are ineffective for other visceral organs, such as liver and pancreas. The introduction of the UW solution in the 1980s, which has proven to be universally effective for the intra-abdominal organs, has been heralded by its inventors as approaching a general organ preservation solution. Moreover, the improvements of UW solution over other organ storage solutions, notably Collins's solution or its variants, have been ascribed to organ-specific differences that demand careful consideration of the choice of components selected to fulfill the objectives summarized above. Nevertheless, UW solution has not proven to be as effective for the intrathoracic organs, heart, and lung as it has for intra-abdominal organs. This is due largely to special demands of highly metabolic tissues such as the myocardium as discussed above.

It is now recognized that the successive phases of the transplantation procedure involving organ procurement, storage, transportation, reimplantation, and reperfusion may impose different requirements for optimum preservation at the different stages. This is illustrated by evidence that heart preservation with the intracellular-type solution, EuroCollins, was enhanced when the heart was initially arrested and subsequently flushed prior to reperfusion, with an extracellular-like cardioplegic solution. It is clear, therefore, that any single formulation of preservation solution is unlikely to provide optimum protection during all the processing stages of a transplantation procedure or the interventional stages of complex surgeries, so a combination of appropriately designed solutions may prove necessary. In addition to the basic elements in the design of cold-storage solutions described above, it has been anticipated that inclusion of additional biochemical and pharmacological agents that might counteract the deleterious effects of cold ischemia and reperfusion injury would enhance the cytoprotection properties of preservation media. This strategy has been explored using either newly designed media that incorporate a cocktail of cytoprotective additives such as Carolina Rinse solution, or by using existing preservation solutions such as UW, or EuroCollins solution, as a vehicle for the pharmacological additives. For example, the addition of aprotinin to UW and EuroCollins solutions has been shown to increase endothelial cell viability in hypoxic cold storage conditions and led to improved lung preservation. The rationale for such studies was based upon the fact that endothelial cell damage, destabilization of mitochondria and cell membranes, and the release of proteolytic enzymes are known to be associated with organ preservation injury, and aprotinin has anti-proteolytic and membrane stabilizing properties.

A wide variety of classes of drugs and biochemical agents have been advocated as potential supplements for improved organ preservation such that complex cocktails of additives might be conceived for optimum organ preservation. Although far from exhaustive, the variety of additives that have been suggested to be potential supplements for preservation solutions but it is outside the scope of this chapter to attempt a discussion or review of these strategies. Nevertheless, one important cautionary point that should be emphasized is the potential interaction of components that might be detrimental, rather than beneficial, to outcome. For example, it has been reported that allopurinol and trifluoperazine improve the function of canine kidneys harvested from non-heart-beating donors when used independently as supplements for Belzer's Machine Preservation Medium. However, when used in combination, these additives unexpectedly proved to be detrimental to preservation outcome and are not recommended for combined use in machine perfusion preservation. Optimization strategies for the design of preservation cocktails must not only take account of the possible interactions of additive components, but also their differential effectiveness at various stages of the procurement, preservation, reperfusion, and reimplantation paradigm. Some studies have indicated that drug administration during hypothermic storage has therapeutic benefits for resuscitating tissues after warm ischemia and is more effective than the same drug given only during reperfusion. Yet other reports indicate that some drugs, particularly antioxidants, are effective only when present in the reperfusion buffer medium, not in the cold-storage solution. Clearly, an unequivocal strategy for hypothermic solution design has not yet emerged and remains a complex issue impacted by many factors including organ-type, condition, pretreatment, storage interval, temperature, and reperfusion conditions.

Multi-organ Protection and Total Body Cooling

Interest in general or universal tissue-preservation techniques is exemplified by the need for methods of protecting multiple vital organs, and even the whole body, for applications in modern-day surgery. Multiple-organ harvesting for transplantation can be optimized by hypothermic perfusion of the whole cadaver, or donor organ blocks comprising several organs, to minimize warm ischemic injury. The ultimate challenge is perhaps protection of the entire body against the effects of global ischemia during periods of circulatory and/or cardiac arrest for "bloodless" surgery.

Surgeons have developed skills that allow very complex, corrective, and life-saving operations to be performed, notably on the heart and brain. As we have reviewed previously, many of these complicated, time-consuming procedures have the inherent need for temporary cessation of blood flow and demand protection of the patient against the deleterious effects of ischemia and anoxia. Although hypothermia is routinely used as an adjunctive protective modality for surgical procedures that require a period of cardiac arrest, there are restrictive time constraints (less than one hour at temperatures usually not lower than 18°C) on the interval of cold ischemia if neurological sequelae are to be avoided. It is well recognized that the window of opportunity for safe surgical intervention could be extended by using greater degrees of hypothermic metabolic suppression, but this becomes unacceptably dangerous, due principally to the effects of profound hypothermia on the blood leading to coagulopathies and irreversible microvascular blockage as discussed above and elsewhere.

The experimental approach we have explored to avoid these complications is to employ a technique of asanguineous blood substitution using acellular synthetic solutions designed to protect the heart, brain, and visceral organs during several hours of bloodless perfusion. The concept of using ultraprofound hypothermia (< 10°C) and complete blood replacement is appealing because deeper hypothermia can provide more effective suppression of metabolism, thereby extending the tolerance to ischemia and minimizing the demand for oxygen to levels that can be adequately supplied in a cold aqueous solution without the need for special oxygen-carrying molecules. Complete exsanguination ameliorates the complications associated with increased viscosity, coagulopathies, and erythrocyte clumping of cooled blood.

Moreover, vascular purging can remove harmful catabolic products and formed elements that might participate in the ischemia and reperfusion injury cascades. Total exsanguination provides the opportunity to control directly the vascular and extracellular compartments with fluids designed to be protective under the conditions of ultraprofound hypothermia. Solutes can be added to maintain ionic and osmotic balance at the cellular and tissue levels; biochemical and pharmacological additives can help sustain tissue integrity in a variety of ways including efficient vascular flushing, membrane stabilization, free-radical scavenging, and providing substrates for the regeneration of high-energy compounds during rewarming and reperfusion. In essence, these are the principles that are embodied to a greater or lesser extent in the design of various solutions used for *ex vivo* organ preservation and we have adopted similar principles in the design of new hypothermic blood substitutes (HBS).

Our working hypothesis has been *that acellular solutions can be designed to act as universal tissue-preservation solutions during several hours of hypothermic whole-body washout involving cardiac arrest, with or without circulatory arrest*. On this basis, we have formulated and evaluated two new types of solutions designated "Purge" and

"Maintenance" HBS that fulfill separate requirements during the asanguineous procedure. The principal solution (e.g., HTS/M or Unisol) is a hyperkalemic intracellular-type solution specifically designed to maintain cellular integrity during the hypothermic interval at the lowest temperature. The second solution (e.g., HTS/P or Unisol-E) is designed to interface between the blood and the maintenance solution during both cooling and warming. This companion solution is therefore an extracellular-type flush solution designed to aid in purging the circulation of blood during cooling since the removal of erythrocytes from the microvasculature is an important objective during ultraprofound hypothermia. The "purge" solution is also designed to flush the system (vasculature and CPB circuit) of the hyperkalemic maintenance solution during warming and possibly help to flush out accumulated toxins and metabolic by-products that might promote oxidative stress and free radical injury upon reperfusion.

Based upon the principles that have emerged from isolated organ preservation studies, an attempt was made to incorporate the important characteristics in the formulation of the hypothermic blood substitute solutions, and components that might fulfill multiple roles were selected wherever possible. Conceptually, this strategy would maximize the intrinsic qualities of the solutions that, by design as universal tissue-preservation solutions, would inevitably be a hybrid of other hypothermic perfusates and storage-media.

These solutions have been shown to protect the brain, heart, and visceral organs during 3.5 hours of cardiac arrest and global ischemia in an asanguineous canine model during controlled profound hypothermia at less than 10°C. More recently, this approach has been applied experimentally to animal models (porcine and canine) of hemorrhagic shock. This novel approach to clinical suspended animation (or *corporoplegia*, meaning literally "*body paralysis*") has been explored for resuscitation after traumatic hemorrhagic shock in preclinical models relevant to both civilian and military applications. In exsanguinating cardiac arrest (CA) conventional resuscitation attempts are futile and result in 100% mortality.

In prior research studies our collaborators at the University of Pittsburgh introduced the use of cold aortic saline flush at the start of CA to rapidly induce protective hypothermia during prolonged CA (120 min) for hemostasis followed by resuscitation. Using a canine model, they showed that a saline flush to a brain temperature of 10°C resulted in normal survival after 90 minutes, but not consistently after CA =

120 minutes. However, an additional study in which Unisol plus the antioxidant Tempol was evaluated as a comparative "optimized flush" showed a markedly improved outcome in regards to physiology, neurology, and brain histology after 120 minutes CA compared with the saline flush.

In separate studies, our collaborators at the Uniformed Services University of the Health Sciences have developed a porcine model of uncontrolled lethal hemorrhage in which a combination of the *Maintenance* and *Purge* solutions were used in a cardiopulmonary bypass (CPB) technique to effect profound hypothermia and prolonged cardiac arrest (60 minutes) with resuscitation after surgical repair of the vascular deficit induced to effect exsanguinations. In the most recent study, after rewarming and discontinuation of CPB, pigs were recovered and monitored for 6 weeks for neurological deficits, cognitive function (learning new skills), and organ dysfunction. Detailed examination of brains was performed at 6 weeks. All the normothermic control animals died, whereas 90% of the HBS animals survived and were neurologically intact, displayed normal learning and memory capability, and had no long-term organ dysfunction. Profound hypothermia markedly diminished total body metabolic activity as evidenced by significantly lower buildup of lactic acid during the periods of hypothermia and rewarming. Histologic examination of brains from the HBS survivors after 6 weeks revealed no ischemic damage in any of the animals, in marked contrast to the brains from control animals, which all showed diffuse ischemic damage.

Successful application of this technique to man would provide a greater than threefold extension of the current limits of less than one hour for "safe" arrest without a high risk of neurological complications. This novel approach to bloodless surgery would significantly broaden the window of opportunity for surgical intervention in a variety of currently inoperable cases, principally in the areas of cardiovascular surgery, neurosurgery, and emergency trauma surgery. This provides further evidence for the protective properties of solutions such as Unisol used for global tissue preservation during whole-body perfusion in which the microvasculature of the heart and brain are especially vulnerable to ischemic injury. Moreover, the application of solution design for clinical suspended animation under conditions of ultraprofound hypothermia places the HBS solutions HypoThermosol and Unisol in a unique category as universal preservation media for all tissues in the body. In contrast, all other preservation media, including the most widely used commercial solutions such as UW-ViaSpan are established for specific organs, or groups of organs (e.g., UW for abdominal organs and Celsior, Cardiosol, or Custodiol for thoracic organs). Moreover, the demonstrated efficacy of these synthetic, acellular, hypothermic blood-substitute solutions justifies their consideration for multiple organ harvesting from cadaveric and heart-beating donors.

Biopreservation employing hypothermic temperatures and appropriately designed solutions has become a critical component of organ transplant therapy during the era of transplantation science. Hypothermic storage of organs is based upon reductions in metabolic demands and oxygen requirements as temperature is reduced. Four levels of hypothermia have been defined in the medical literature as mild (32° to 35°C), moderate (27° to 32°C), deep or profound (10° to 27°C), and ultraprofound (< 10°C). Profound hypothermia has been the most common hypothermic temperature range applied for tissue and organ preservation. Initially, hypothermia is usually applied by flushing organs with cold solution either just prior to removal from the donor or immediately after removal from the donor. The solution is left in the organ vasculature and surrounds the organ during hypothermic storage and transportation to the recipient. In some cases, the solution bathes the exterior of the organ and is actively perfused through the organ's vasculature during hypothermic storage and transport. Traditionally, different solutions are used for flushing and *static* hypothermic storage of organs than are used for *perfusion* hypothermic storage of organs as outlined above.

For cell and tissue preservation the history of preservation strategies, particularly with regard to appropriate solution design, is not so well established. There have been very few studies aimed at optimizing solution design specifically for isolated cells, tissues, or engineered tissue constructs. In some cases hypothermic organ preservation solutions have been used, but in general biologists have tended to stick with what they know and have used tissue culture media, even for low-temperature exposure. For the reasons expounded in this chapter this is inappropriate and it is preferable to focus on the formulation of new solutions designed to protect cells and tissues during hypothermic exposure. Similarly for cryopreservation the role of the vehicle solution for cryoprotective additives (CPAs) as a determinant of cell survival has often been overlooked, or trivialized, by assuming that regular tissue culture medium will suffice. There is now substantial evidence that this is not the case and a strategy to design new solutions that take account of the biology of cell survival in the cold clearly leads to improved methods of biopreservation.

8

Organ Preservation

The ability to preserve organs outside the body has played a major role in the development of transplant services worldwide over the last 4 decades. Indeed, although the recognition and control of immune responses elicited by the allograft reaction have been central to the success of transplant therapies, equal recognition must be given to preservation technologies that have permitted the retrieval and transport of organs between hospitals in different cities and sometimes in different countries. As stated by Calne and colleagues in the opening line of their report in 1963, "Ischaemic renal damage is extremely important to those interested in cadaveric kidney transplantation and has probably been responsible for more failures than immunological reactions." Without the development of modern preservation protocols, organ transplantation would have remained a very esoteric and infrequent operation. This chapter reviews the history and development of organ preservation, the current status of protocols used in the major transplantable organs, and future avenues that may be pursued.

History of Organ Transplantation

The scientific challenge of maintaining a functioning organ outside the body has a long history and has been inextricably linked with the expansion of physiology and medicine in the 19th and 20th centuries. An understanding of the essential support provided by the circulatory system fueled attempts to perfuse *in vitro* organs such as the kidney and liver by simple syringe methods, whereas the development of more sophisticated perfusion circuits using artificial pumps allowed detailed physiology of organ function to be studied outside the body. In parallel, investigators started to formulate synthetic perfusates-electrolyte solutions capable of replacing blood, at least in some respects. By the 1930s, sufficient significant progress had been made to permit Carrel and Lindbergh to study a range of organs *in vitro*, including kidney, heart, lung, spleen, thyroid, ovary, and fallopian tube at normal body temperatures for several days. Despite this success, these pioneers recognized the limitations of their systems to maintain organ viability without associated perfusion damage (resulting from the inability to completely mimic all the essential support provided by the *in vivo* circulatory system), most often indicated by increasing edema (now known to result from failing microcirculation) and associated hypoxic damage. Thus, an additional safety net for the maintenance of organ viability was sought; this solution proved to be the application of hypothermia.

Although sporadic attempts had been made to transplant organs using microsurgical techniques throughout the early part of the 20th century, it was not until the 1950s that consistent operative success was achieved by Hume and colleagues, most frequently in the kidney. These authors themselves commented on the problems of ischaemic damage to the grafted kidneys and on the negative effect this damage had on postoperative renal function. In the same period, there were coincident reports about the effects of cooling on renal physiology, and others on the protection provided by hypothermia and afforded to kidneys *in vivo* during ischemia. Thus, there were many logical reasons for using hypothermia in organ preservation *in vitro*, and these reasons were applied by Calne et al. to demonstrate that protection of function to the grafted kidney could be achieved. These authors investigated the relative merits of cooling kidneys by simple surface cooling or by perfusion of the renal artery with cooled heparinized blood and concluded that vascular perfusion was more efficient. From these and other early studies, the basic tenets of organ preservation were established (i.e., to perfuse via the vascular bed with a specially selected chilled solution). It was soon established that heparinized or diluted blood, when cooled and used as preservation solution, still led to problems with vascular stasis on reimplantation of the graft, and the next phase of research focused on better solutions to be used for the establishment of *in vitro* hypothermia.

Organ Preservation: The Problems of Cooling and Hypoxia

In mammals, up to 70% of basal energy metabolism is concerned with the maintenance of homeostasis, on which all the complex, integrated life processes depend. The control of ion distribution, pH, solute content, and associated osmotic potential within cells all depend on this high-level basal activity, particularly for fueling membrane pumps such as the Na^+/K^+ ATPase. Such significant energy requirements can only be met by oxidative energy production via coupled mitochondrial electron transfer, and thus it is unsurprising that transplantation (requiring removal and manipulation of organs) has such a propensity to induce hypoxic damage.

The inescapable need to use energy (in mammalian systems in the form of ATP and phosphocreatine) for homeostasis means that once vascular perfusion with blood is interrupted, ATP levels at normal body temperature fall dramatically within minutes and reach a low plateau value over approximately half an hour. Cells possess various fallback mechanisms to generate energy by hypoxic metabolism, such as anaerobic glycolysis and limited substrate interconversion, but these mechanisms can only provide a small fraction of the required energy supply and can only function for limited periods. Thereafter, the homeostatic controls begin to differentially "unravel," leading to progressively more injury, which at some point becomes irredeemable, even if the organ is transplanted and blood perfusion reestablished. There has been continuing debate over the last 40 years about which injury is the key factor, but in reality, this factor is an interlinked cascade of hypoxic changes, the sum of which cannot be controlled by any simple means such as administering pharmacological agents.

The issue is further complicated because this cascade itself predisposes the cells to additional damage when oxygen is supplied during reperfusion of the transplanted organ, leading to the classically defined ischaemia-reperfusion syndrome. It is beyond the scope of this chapter to describe this syndrome, and several recent reviews have covered this field in detail. It is clear that inhibition of oxidative energy supply (both by deprivation of essential ATP for transmembrane pumping and by structural disruption of the mitochondria that signals secondary damage such as initiation of apoptosis) is one of the most significant factors in the scheme.

The use of cooling to try and control the damaging effects of organ manipulation outside the body is, of course, intuitive, as man has used low-temperature techniques such as ice-packing to slow biodeterioration in foods and crops for thousands of years. Also, many microscopic studies undertaken on cells at reduced temperatures documented the associated decrease in activities, such as motility and streaming of cytoplasm. In current practice, hypothermia in organ preservation is defined as storage at any temperature down to 0°C in the liquid state.

The empirical studies in the middle of the last century demonstrated that mammalian systems could be cooled by hypothermia and then recover function. However, when viewed considering the complexity at normal body temperatures of the multitude of coupled biochemical processes essential for life, it is sometimes difficult to understand how that temperature can be lowered by tens of degrees and then returned to normal without devastating consequences. The secret may partly lie in evolutionary development. Homeotherms may have evolved from earlier ancestors in which body temperature was allowed to fluctuate more widely as the environment changed; this theory is backed by the observation that some of the more primitive current mammalian species (monotremes and marsupials) function at lower and more widely ranging body temperatures. It is also true that the vestigial ability to allow body temperature to fluctuate widely in cold winters in specialized groups of hibernatory mammals indicates that there may be an evolutionary route to the ability to withstand hypothermia, but this is a highly specialized response, in which weeks of metabolic preparation with allied changes in feeding practices, hormone profiles, and molecular and physiological adaptations are necessary for survival at low temperatures. This process is vastly different from the immediate requirement for cooling in transplantation, which is one reason why it has been difficult to translate research on hibernation into applied hypothermia in organ preservation.

The physiological basis for the depression of metabolism was established by the work of Arrhenius and modified for biological processes by authors such as Heilbrunn. Many important biochemical reactions require prior activation of the participating molecules; that is, molecules possessing sufficient energy to take part in the process (this is why cellular energy balance is so important in all life processes). Cooling (by removing energy in the form of heat) reduces the proportion of activated molecules, which can be expressed in the relationship between reaction rate (V) and temperature (T, in degrees absolute), where R is the gas constant, E_a is the activation energy and A a constant:

$$V = A^{\exp} (E_a/RT).$$

Results of a specific metabolic reaction during cooling, when plotted as Logn V against $1/T$ provide a linear relationship with a slope of E_a/R. This can be used to give the activation energy of the specific reaction (assuming a single, rate-limiting step). Over the temperature range used for organ preservation (from body temperature to 0°C), many biochemical processes in mammalian systems do not produce a single linear Arrhenius relationship on cooling but, rather, express a curvilinear or "broken" plot, with a shift to higher activation energies (i.e., a slower overall rate) at temperatures below about 20°C. The depression caused by cooling can also be quantified as the Q_{10} relationship, in which the fall in reaction rate for a 10°C temperature drop is calculated. For many biochemical reactions, the Q_{10} between 40°C and 20°C is about 2, whereas between 20°C and 0°C the value increases to between 3 and 5. Taken together, these observations have been interpreted as the increasing and additive effects of cooling on cellular ultrastructure that affect biochemical processes. (For example, many important biological reactions are catalyzed by enzymes embedded in the membrane lipid bilayers or require substrate exchange across such bilayers.)

It should be stressed here that most of these physiological studies on cooling were performed on systems that were aerobically respiring. The complicating effects of hypoxia (as experienced by organs removed from the body for transplantation) can be seen as an additional metabolic challenge. In simple terms, hypothermia can be aerobic or anaerobic; even under aerobic conditions there is a depression of all metabolic activity, and true, long-term adaptation to cold (as in hibernation) requires a level of complexity far greater than simply applying low temperatures. It is this deficit between the aspirations of the transplant surgeon and the ability of the organ to survive the storage period that has been the driving force for the modern studies in organ preservation.

Cooling and Cellular Energetics

Mitochondrial electron transport and ATP production are surprisingly robust in mammalian cells after cooling. For example, it has been known for some time that in kidneys subjected to oxygenated hypothermic perfusion for up to 48 hours, both the total content of adenine nucleotides and relative ratios of ATP to lower phosphorylated-state adenine nucleotides could be maintained unchanged compared to those measured in kidneys at normal body temperatures. This is true metabolic turnover, rather than a static maintenance of an existing pool, as indicated by experiments on liver. In organs previously depleted of ATP by cold hypoxic storage, a progressive restoration of ATP over several minutes could be shown when oxygenated perfusion was resumed. It has also been known for some time that ATP is consumed at hypothermia for homeostatic ion pumping, as addition of uncoupling agents in stable, oxygenated organs caused a massive release of intracellular potassium. Similar effects of ion-transport inhibitors on energy consumption have been demonstrated in the hypothermic kidney. It is necessary to recognize that measurements of cell energy supply are always a balance between production and consumption, such that the "normal" status measured in these organs can equally be attributed to a lowered energy consumption at hypothermia, which is just balanced by a lower (but adequate) activity of mitochondrial electron transport. Other methods of oxygen supply at hypothermia (such as persufflation) will be discussed later and have similar positive effects on energy turnover. However, the central tenet holds that oxygen is still essential, even at cold temperatures.

Once oxygen is restricted, in the cold, flush-stored organ, cooling itself provides minimal protection of energy balance. The switch to anaerobic metabolism via glycolysis produces low yields of ATP, which at normal body temperature can in no way match the requirements for continued homeostatic control, such that, within a few minutes, ATP falls to very low levels. Cooling, by the relative Q_{10} depression of all metabolic processes, delays this energy collapse by a factor of about 8, or put in another way, adenine nucleotide status fell within about 240 minutes to the equivalent of about 30 minutes at 37°C.

Although a continued energy supply can permit many of the essential homeostatic controls of ion gradients and pH, it is easy to foresee that other, more complicated multistep processes such as protein synthesis may be more susceptible to cooling; this is, in fact, the case. It is difficult to detect synthesis of new proteins in response to a specific inducer much below about 37°C.

Effects of Cooling on Cell Structure

All essential cell metabolism depends on the separation of discrete chemical reactions into specific compartments by the selective properties of biological membranes (comprising phospholipid bilayers and proteins in a liquid crystalline or fluid state at normal body temperatures). Lipids generally are known to undergo a phase transition to a gel state on cooling, and in the complex mixtures of lipid classes in biological membranes, cooling induces a thermotropic separation of the different lipid classes as the respective melting point of a particular lipid is passed. This leads to demixing and lateral phase separation into gel and liquid crystalline phases in the plane of the membrane, and also an exclusion of integral proteins from the gel-phase areas. Such gel formation acts to decrease global membrane fluidity and may also lead to increased solute leakiness through membranes in which such packing faults have occurred. Cooling has been reported to increase transmembrane diffusion of many solutes from small ions to larger molecules, such as disaccharide sugars. The change in fluidity affects the kinetics of many membrane-bound enzymes, such as the Na^{+}/K^{+} ATPase, in which it was shown that cooling of the enzyme when bound in the membrane had a greater inhibitory effect than when the enzyme was a free protein in solution.

Cooling has the potential to alter the cytoskeletal arrangement in mammalian cells. The maintenance of internal structures such as microtubules or microfilaments relies on a sensitive interplay of ion balances, energy status and temperature. In isolated hepatocytes, microtubules were shown to de-polymerize and F-actin from microfilaments to precipitate during storage at 4°C. Alteration in the colloid content of the preservation solution could beneficially modulate these changes. A similar change in hepatocyte microtubules was reported by Kim & Southard, which again could be modified by drug action. In kidney cells, a cold-induced disruption of microtubules was also observed, and was associated with a redistribution of integral membrane proteins.

Mitochondria are also sensitive to reduced temperatures during organ preservation, and swelling or vacuolization has been demonstrated by electron microscopy in experiments on kidney and liver storage. Membrane reorganization with loss of electron chain complexes has also been reported.

Calcium, which is normally sequestered in both endoplasmic reticulum and mitochondria, is also known to be released, significantly increasing cytosolic free-calcium concentrations. Particularly during cold hypoxic storage, conditions prevail that may stimulate the opening of the mitochondrial transition pore, which may result in release of cytochrome C and stimulation of cell death via apoptosis, particularly during the rewarming phase after preservation.

Cooling may also affect the molecular organization at the level of specific proteins. Some enzymes, notably those that are multisubunit enzymes, are susceptible to cold-related inactivation by dissociation. The pK of specific ionizable groups of proteins may alter with lowered temperature, which affects associated reactions within and between proteins. There is also a noted shift in the pK of the most common

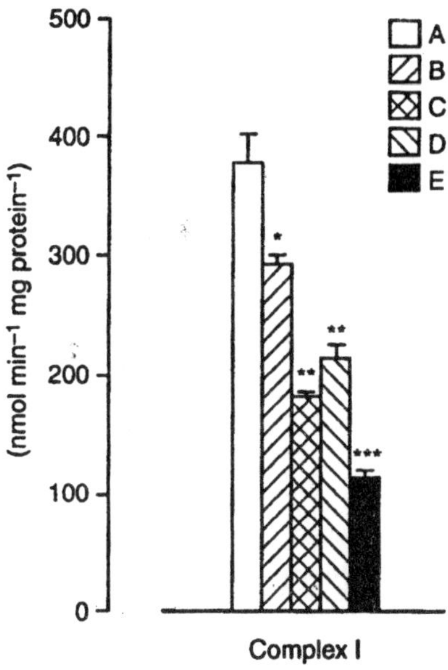

Fig. 8.1. Hypothermic storage and transplantation and the activity of hepatic mitochondrial complex I.

protein-associated ionizable groups, (the -imidazole group of peptide histidine residues), which characteristically appears as an apparent shift in pH (by a factor of about 0.017 pH unit/°C, on cooling from 37°C). This is not a true alkalinization, because the effect is to maintain the relative degree of ionization of the α-imidazole pool, which is the major intracellular protein-associated buffer. The apparent shift in pH can be demonstrated by techniques such as whole-organ nuclear magnetic resonance spectroscopy and is evident even in those organs in which energy metabolism is supported by oxygenated hypothermic perfusion. [48] Other changes may involve proteins associated with binding transition metals, particularly iron. There is growing evidence that iron-associated free radical effects are an important source of cold-preservation damage. Normally, non-protein-associated chelatable pools of iron are very low, but following hypoxic cold preservation, this "free" iron has been shown to increase. This has been suggested to play a role in apoptosis and cell death.

The destabilization of homeostasis when hypoxia is coupled with cooling leads to a predictably complex cascade of biochemical events that eventually become irreversible. Progressive dephosphorylation of ATP results in accumulation of breakdown products such as purines and nucleoside bases (e.g., inosine, hypoxanthine, and xanthine), which

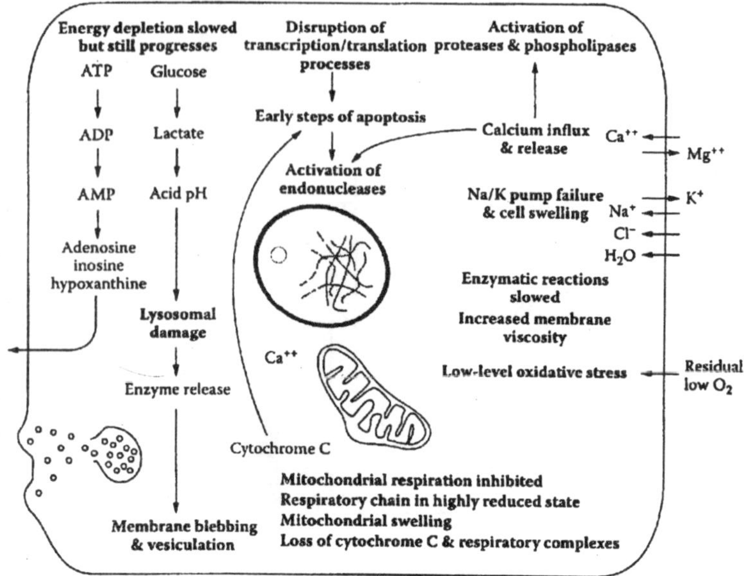

Fig. 8.2. A schematic of a cell during cold hypoxia.

can diffuse from the cell. The internal pH falls toward acid conditions as the switch to anaerobic glycolysis fuels accumulation of lactate. Failure of energy-linked transmembrane pumps results in an inward flux of Na^+ (enhanced by resident anions on intracellular proteins) and Cl^-, leading to associated water influx and cell swelling. Phosphate levels increase as a result of the dephosphorylation events, and control of other actively transported ions (such as calcium and magnesium) is lost. The high-free calcium levels can cause mitochondrial damage and activate degradative enzymes (such as phospholipases and nonlysosomal proteases), which destabilize membranes, accelerating the abnormal leakage of solutes. Taken together with the lowered pH, which also favors the activation of lysosomal digestive enzymes by destabilizing lysosomal membranes, an autodigestive state is enhanced. Depending on the time and conditions experienced during cold hypoxia, progressive cell death by both necrosis and apoptosis has been observed. The endothelial cell compartment of many organs is particularly sensitive to cold hypoxic preservation, leading to membrane blebbing, cell detachment, and blockage of the microcirculation. Residual blood cells within the vascular bed also become swollen and act to plug the capillary network. This extensive biochemical understanding is beginning to be supplemented by molecular studies that indicate that cooling itself may activate specific signaling pathways that can drive transcription of selected genes on rewarming.

EFFECTS OF COOLING: ENHANCING REPERFUSION INJURY

The multitude of changes described above that occur during the cold storage period (especially in the hypoxic state) predispose the organ to further injury during the early rewarming phase. These additional events have been collectively termed reperfusion injury or ischaemia/reperfusion injury (IR). The syndrome has been recognized since the early days of organ preservation and was then called the "no-reflow phenomenum." A great deal of information on the driving forces for the development of IR injury (which occurs after both warm and cold hypoxia) has been accumulated over the past decade, and a full description is beyond this scope of this chapter, but it is worth summarizing some of the major points.

Many of the mechanisms are central to all situations, but individual factors (such as calcium- induced contracture of cardiac muscle) are particularly obvious in specific organs. Mitochondria are as important in IR injury as they are in preservation injury. The conditions during reperfusion can favor a more substantial and prolonged opening of the mitochondrial transition pore following loss of calcium homeostasis, leading to apoptosis or impaired electron transport and failure of ATP regeneration. Impaired mitochondrial function also leads to inefficient transfer of electrons and generation of oxygen free radicals when oxygen is resupplied on revascularization, adding to a generalized oxidative stress from multiple causes.

Delocalization of catalytically available free iron can enhance *oxygen free radical* (OFR) formation. One of the earliest proposals for OFR production during reperfusion injury was that of the breakdown products of adenine nucleotides, particularly to xanthine and hypoxanthine, interacting with the xanthine oxidase enzyme to liberate OFR, and this scenario is still being evaluated today. Endothelial cells within the vascular bed of the reperfusing organ will generate nitric oxide (itself a radical) as part of the physiological control of vascular tone, and this additional radical may interact with other OFR to produce harmful adducts such as peroxynitrite. In specific situations, cell subsets (such as Kupffer cells in the liver or mesangial cells in the kidney) can be activated by IR to release additional sources of OFR or contribute directly to the inflammatory response, in addition to effects caused by influx of neutrophils within the recirculating blood. In fact, a secondary sequence of changes results from interaction between the

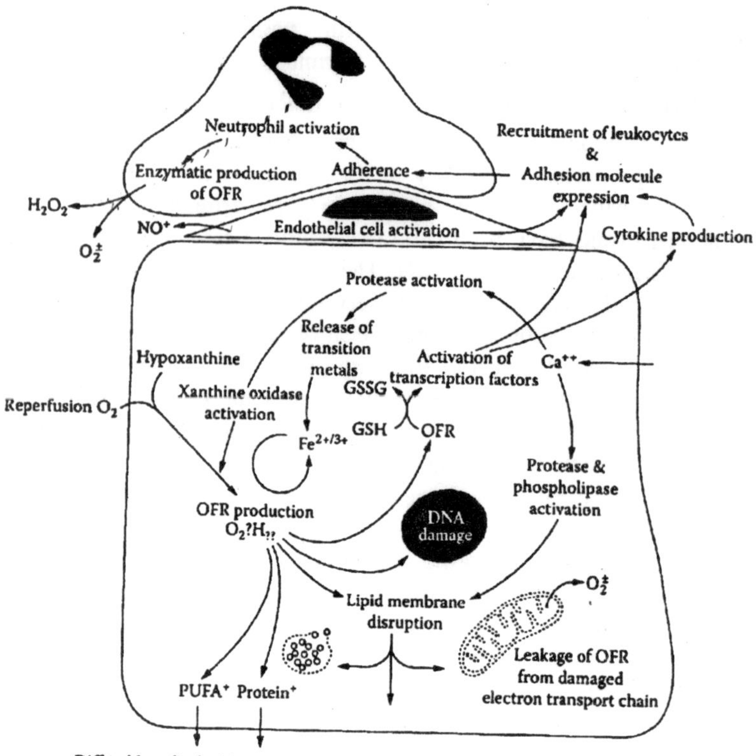

Fig. 8.3. A schematic of the pathways for oxygen free radical (OFR) production following cold hypoxic preservation.

damaged organ and the host's blood system during reperfusion. The variable redox status of the organ at this time can activate various transcription factors such as hypoxia-inducible factor (HIF-1) and nuclear factor kappa B, which control production of inflammatory markers. The effects can also enhance local synthesis of cytokines and chemokines, attracting leukocytes to the organ. The same agents activate expression of cell adhesion molecules on the cells of the reperfusing organ, which increases binding of leukocytes, particularly neutrophils, and magnifies the inflammatory response. Neutrophils themselves produce OFR and other agents such as proteases, which can enhance the degree of IR, with additional stimulus for platelet activation resulting in blockage of the microcirculation.

This combination of preservation- and reperfusion-related injuries can have negative effects on graft function well beyond the immediate post-storage period. There is increasing evidence that the stimulation of the IR inflammatory cascade may produce a prolonged activation of the host's immune system (beyond that produced directly by the allograft reaction to the foreign organ), producing progressive fibrotic change, which eventually results in deteriorating graft function or loss. These consequences explain why so much effort over the last 2 decades has been placed on developing preservation solutions, with pharmacological additives, to ameliorate the damage.

A Comparison to Natural Cold and Hypoxic Survival

Although hibernation and natural cold tolerance are complex fields too detailed to discuss in depth here, there are a few points that are worth making in comparison to the aims of organ preservation techniques. As discussed above, the maintenance of cellular homeostasis is as important for the survival of cold- and hypoxia-tolerant species as it is to normal homeotherms. Across the spectrum of species that survive in extreme environments (from lower vertebrates such as turtles to true hibernators such as ground squirrels on the North American continent), the necessity is to balance energy consumption with production under conditions in which oxygen and nutrients may be limiting. This requires a range of responses that have been defined as defense and rescue.

There are integrated gene-regulated responses that down-regulate energy consumption in the resting condition to a new hypometabolic state. The changes range from alterations in membrane channel activity in individual cells to a whole-body response from pH change. Although many lower species (such as turtle) have a relatively efficient capability to harness glycolysis and other secondary fermentations, it is instructive to note that the majority of mammalian hibernators strive to maintain oxidative metabolism (albeit at a low rate) — they perform intermittent breathing motions while in deep hibernation. It has also been found that many hibernators go through several brief arousal periods during the hibernation season that could indicate a need for the rescue of essential metabolic pathways required for the essential homeostasis. In all species there appears the need for the maintenance of a small, consolidated, turnover of ATP — energy metabolism cannot be completely silenced even in the best-adapted species but must be maintained at a low "*pilot light*" level of ATP regeneration. Given these intriguing responses, there may be lessons to be learned from hibernating species in applied organ preservation. Some work has been performed in this area, but much remains to be understood.

Preservation Strategies

The sudden arrival of clinical kidney transplantation in 1962–1963 was so unexpected that little collateral research or other formal preparation had been made to preserve organs. Although kidneys were successfully transplanted in the pioneer identical twin cases despite protracted periods of warm ischemia, the maturation of clinical transplantation could not proceed without effective organ conservation. This conservation was at first accomplished by total body hypothermia of living volunteer kidney donors, using methods developed by cardiac surgeons for open heart operations. In the experimental laboratory, Lillehei et al. simply immersed the excised intestine in iced saline before its autotransplantation, a method also used by Shumway when developing experimental and clinical heart and heart–lung transplantation. Thus the principle of hypothermia was understood at an early time, although not efficiently applied.

The first major innovation in hypothermia was in the laboratory when canine liver allografts were cooled through the infusion of chilled fluids into the vascular bed of hepatic allografts via the portal vein. Before this time, after liver transplantation, dogs almost never survived, whereas after this technical innovation, success became routine. In a logical extension of this knowledge to clinical kidney transplantation, in 1963 the practice of infusing chilled lactated Ringer's or low-molecular-weight dextran solutions into the renal artery of kidney grafts immediately after their removal was introduced. The solution described by Collins, Bravo-Shugarman, and Terasaki, resembling intracellular electrolyte concentrations, or modifications of it, was the first notable advance in organ preservation and was used for almost 2 decades.

Renal allograft preservation was feasible for 1–2 days, long enough to allow tissue matching and sharing of organs over a wide geographic area. Experiments with hepatic allografts by Benichou et al. using the Collins-Terasaki solution, and by Wall et al. with the plasma-like Schalm solution, led directly to liver sharing among cities, but with a time limitation of only 2–8 hours.

Today, intravascular cooling is the first step in the preservation of all whole-organ grafts. For cadaver donors, this is most often done *in situ* by some variant of the technique described originally by Marchioro et al. Ackermann and Snell and Merkel, Jonasson, and Bergan popularized the simpler core cooling of cadavers with cold electrolyte solutions infused into the distal aorta.

Until 1981, transplantation of the extrarenal organs was an unusual event. By late 1981, however, it had become obvious that liver and thoracic organ-transplant procedures were going to be widely used. A method of multiple-organ procurement was required by which the kidneys, liver, heart, and lungs — or various combinations of these organs — could be removed without jeopardizing any of the individual organs. Flexible techniques were developed that were quickly adopted worldwide. With these methods, all organs to be transplanted are cooled *in situ*, rapidly removed in a bloodless field, and dissected on a back table.

Fluids of differing osmotic, oncotic, and electrolyte composition are infused into the allograft before placing it in a refrigerated container. Introduction of the University of Wisconsin solution (UW) to liver transplantation by Belzer, Jamieson, and Kalayoglu was the next major development in static preservation. The superiority of the UW for preservation of the kidney and other organs was promptly demonstrated in experimental models and confirmed in clinical trials. The UW preservation doubled or tripled the time of safe preservation of the various allografts, making national and international sharing of most organs an economic and practical objective.

From the above discussions, it will have become apparent that there is a fundamental choice to be made in applied organ preservation: whether to supply oxygen to sustain low-level oxidative metabolism and energy-consuming homeostatic control of the intracellular milieu or whether to allow hypoxia to develop and attempt to modulate the inevitable biochemical changes such that organ viability is maintained for a practically useful period. Hypothermia is central to both approaches, but thereafter they differ in complexity, portability, and cost. There are some general approaches for correcting hypothermic damage, and these approaches will be outlined here. Specific compositions of preservation solutions for static or perfusion preservation will be discussed later in this chapter.

Maintaining Homeostasis in Preservation

Oxygenation

Static cold storage cannot supply oxygen, whereas continuous organ perfusion with an asanguinous-oxygenated perfusion can provide adequate oxygen for a hypothermic organ's low metabolic requirements. In the kidney, perfusion with oxygenated Fluosol may improve preservation further, whereas retrograde oxygen persufflation has proven to be a useful adjunct to static preservation in experimental studies but has not yet gained wide acceptance because of the fear of increased endothelial vascular damage.

Nutrition

Glucose is often incorporated as a metabolic fuel, but amino acids, fatty acids, ATP, and adenosine have also been added, particularly in

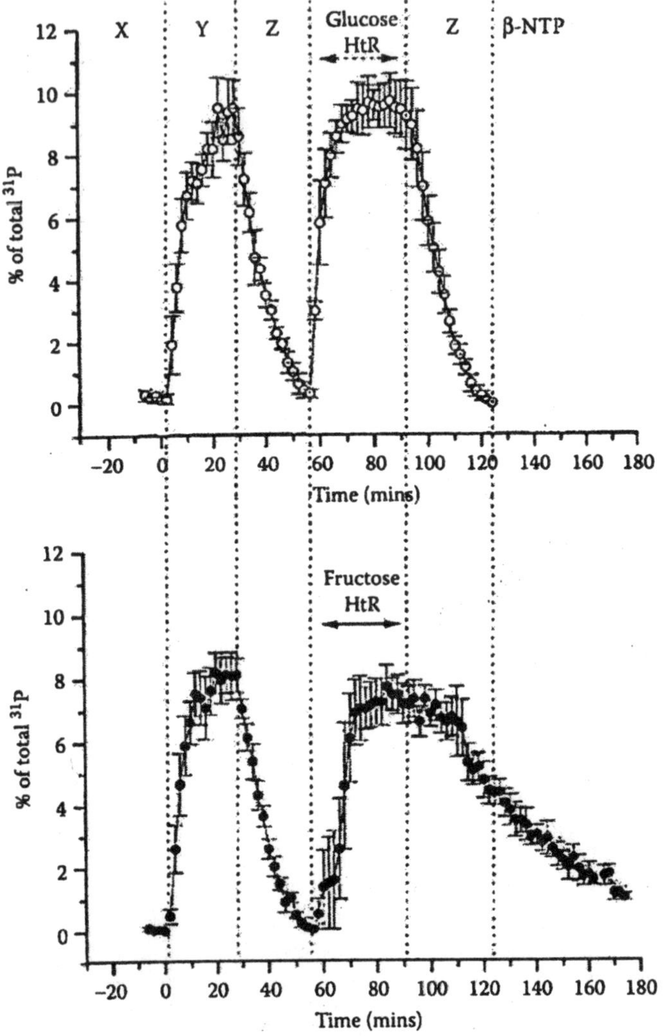

Fig. 8.4. Real-time changes in hepatic NTP following liver isolation and reperfusion with glucose and fructose substrates.

perfusion preservation. Complete oxidative phosphorylation may require such additional supplements during perfusion. Adenosine was found to be an essential additive in the UW for static storage, although its vasoactive properties may also have facilitated the flushing procedure. During hypoxic storage, provision of substrates such as fructose has been shown in experimental studies to prolong the synthesis of ATP for a limited period.

pH maintenance

Flushing solutions with good buffering capacities are generally used for static cold storage (e.g., phosphate, citrate, and histidine). Bicarbonate buffering, which is generally used for continuous perfusion, requires continuous provision of carbon dioxide and is not as effective in the pH range 7.1–7.8, but protein in the perfusate can offer additional buffering capacity. Other components of static flush solutions such as citrate, phosphate, and lactobionate have limited buffering capabilities. Histidine is an efficient buffer at low temperatures and forms the basis of Bretschneider's solution, now widely used for static preservation of kidneys in Europe. The addition of glucose to solutions for static preservation may be counterproductive if lactate production is enhanced and acidosis develops. Such additions for continuous perfusions can prove effective because acid catabolites are continually diluted and washed out of the organ.

Prevention of cellular edema

The use of solutions containing slowly permeable (e.g., mannitol and glucose) or impermeable (e.g., sucrose and raffinose) solutes helps reduce edema during both static and perfusion storage. Further advantage can be obtained by replacing the permeable anion, chloride, with an impermeable anion (e.g., lactobionate or gluconate). The gradients affecting loss of potassium and magnesium and influx of sodium can be minimized by preservation solutions containing high intracellular concentrations of potassium and magnesium and low concentrations of sodium. However, sodium/potassium ratios for static and perfusion storage can be reversed without significant ill effect.

Modification of cellular calcium

Diltiazem, a calcium antagonist, has attracted attention because it increases renal blood flow and prevents an influx of calcium. In the kidney, diltiazem has been shown to reduce the incidence of *acute tubular necrosis* (ATN) when incorporated in EuroCollins' (EC) flush-out solution, administered during kidney procurement and after transplantation. Another calcium antagonist, verapamil, has similar protective properties. Incorporation of phenothiazines such as trifluoperazine into flushing solutions has also been used to inhibit the activation of the calcium–calmodulin complex and has improved kidney preservation.

Antioxidants

Several strategies have been tested to limit free-radical formation after reperfusion of the graft. Allopurinol, a xanthine oxidase inhibitor, has been added to the preserving solution or administered to the donor and recipient, with inconclusive results. Its omission from UW was found to be deleterious to renal function. Free-radical scavengers glutathione and ascorbate have also been included in some preservation solutions, but the efficacy of such additives may be limited by their instability in the presence of atmospheric oxygen during production or storage of the solution. Mannitol, polyethylene glycol, histidine, and tryptophane have also been used. The iron-chelating agent deferoxamine has been an effective additive to the perfusate for machine preservation of canine kidneys for 48 hours with 30 minutes of warmischaemia (WI).

The effect of hypothermia on enzymes and membranes cannot easily be manipulated. In nature, such modification is a characteristic of hibernating animals that can withstand low temperatures for long periods with no damage. Kidneys from hibernating squirrels can be preserved for 10 days and maintain metabolic activity comparable to that measured after only 3 days from nonhibernating squirrels.

Donor Factors in Organ Preservation

Although the main thrust of this chapter has been the changes imposed on individual organs by cold preservation, there are some factors inherent in the donor from which the organs have been removed that affect storage outcome.

Brain death

The primary source of solid-organ transplants has become brain-dead cadaveric donors. Although brain death has been well defined neurologically, knowledge of the systemic changes that follow this massive central injury remains limited. Growing interest in the pathophysiology of the condition and its effects on function of peripheral organs has led to a number of experimental and clinical studies designed to elucidate the complexity of hemodynamic, neurohumoral, and immunologic alterations that may develop. The first major studies of the effects of brain death on the structure and function of organs to be transplanted were carried out by Wicomb et al. in the early 1980s, stimulated by an observation made in the first patient to receive a donor heart that had been stored by continuous hypothermic perfusion. After heterotopic transplantation, this heart, which had been hypothermically perfused and stored for over 12 hours, did not function for almost 24 hours, in direct contrast to observations made in baboons receiving orthotopic transplants of hearts that had been stored under identical conditions for periods of 24 hours. The significant difference between these clinical and experimental studies was that although in the baboons the donor hearts were excised from healthy, anesthetized animals, the human heart had been obtained from a brain- dead donor. In numerous studies in both baboons and pigs, the Cape Town group confirmed and expanded on Cushing's observations (made in 1901) on the hemodynamic consequences of an increase in intracranial pressure, often associated with brain death.

Inflation of the balloon of a Foley catheter (placed within the skull) resulted in the typical features of the "*autonomic storm*," which have been fully described previously. This intense stimulation of the autonomic nervous system resulted in major electrocardiographic and hemodynamic abnormalities and even in structural damage to the myocardium and lungs. Novitzky's group also documented significant reductions in the circulating levels of certain hormones, notably free triiodothyronine (fT3), cortisol, and insulin, which in turn led to major metabolic disturbances. Anaerobic metabolism replaced aerobic metabolism, myocardial high-energy stores were depleted, lactate was accumulated, and organ function deteriorated.

Brain death involves a syndrome that includes rapid swings in blood pressure, hypotension, coagulopathies, pulmonary changes, hypothermia, and electrolyte aberrations. The hemodynamic changes are well recognized. Hypotension following brain death is a common phenomenon necessitating intensive stabilization. It has been confirmed in rats that brain death produces noradrenalin levels that are 22-fold and adrenalin levels that are 148-fold of baseline. This increase in catecholamines was followed by altered organ perfusion secondary to elevated vascular resistance (kidney fourfold, spleen ninefold, liver twofold to those resistance levels seen in control conditions). Fifteen minutes after central injury, the mean blood pressure and vascular resistance decreased below normal, indicating increasing anoxia. Vasoconstriction during the initial events is severe, and a significant reduction in flow to peripheral organs occurs despite the highly increased perfusion pressure; the combination results in ischemia. Elevated catecholamine levels secondary to brain death appear to cause major histopathologic and functional effects in organs, best identified in the heart. Cardiac lesions include petechial hemorrhage in the subendocardium, contraction bands, and coagulative myocytolysis, which is also found in patients dying of acute cerebral injury as well as in experimental animals following catecholamine administration.

The type of brain death appears to influence subsequent events. The effects of brain death are less severe following brain death caused by a gradual increase in intracranial pressure, where there is a transient decrease in the heart rate but no changes in other hemodynamic parameters and less severe morphologic changes of heart tissue. This variable course of hemodynamic instability appears to be the result of the extent of brain stem lesions. These differences are emphasized by multivariate analysis, which shows that both primary and long-term results of organ transplantation are dependent on the etiology of brain death in the donor. The prognosis for graft success is poorest when the donor has died of acute intracerebral hemorrhage, where there is an explosive increase of intracranial pressure and hemodynamic instablity. In addition, administration of exogenous catecholamines to brain-dead donors is followed by less satisfactory survival and initial graft function after allotransplantation.

Little is know about the effects of brain death on cytokine activation of peripheral organs. Examination of the presence of cytokines in the circulation of brain-dead patients showed increased interleukin 6 levels and up-regulation of transcriptional levels of tumor necrosis factor alpha (TNF-α) and interleukin 6 in association with episodes of transient focal cerebral ischemia. It has been suggested that organs experiencing warm ischemia because of hemodynamic deterioration of brain-dead donors, and subsequently cold ischemia during storage, express higher levels of cytokines than those from ideal living donors by the time they are transplanted. Recent experimental studies investigating the relation between brain death and the activation of peripheral organs have demonstrated that an explosive increase in intracranial pressure in rats rapidly upregulates various lymphocyte and macrophage-derived cytokines on peripheral organs. Up-regulation of MHC class II and I complexes and the costimulatory molecule B7 indicates that the immunogenicity of injured organs was increased.

The hypothesis that brain death increases the immunogenicity of solid organs is further supported by clinical findings that indicate that the incidence and severity of acute rejection are greater among cadaveric organs than among those from living donor sources. This would explain the apparent synergy noted clinically between the effects of initial delayed graft function and acute rejection episodes. Whether brain death–associated activation of donor organs is mediated via neurohormonal, circulating cytokines or by hemodynamic events following central destruction remains unanswered.

Warm ischemic time and non-heart-beating donors

The standard organ retrieval procedure from cadaveric donors diagnosed with brain death, but still maintaining a cardiac circulation, produces minimal warm ischaemia before the organ cooling and preservation. However, the growing imbalance between organs available for transplantation and needy patients has led to consideration of using organs from less favorable situations — the so-called marginal donors. Within this group are donors who have gone into sudden cardiac arrest and those in whom supportive therapy has been withdrawn on ethical grounds but who do not fulfill brain-death criteria. In these situations, considerable agonal ischemia (up to 1 hour) may be experienced before organs can be surgically removed and cooled — this group has become known as the non-heart-beating donor pool. It is beyond the scope of this chapter to discuss these issues, but information can be found in several recent publications. However, there is considerable evidence that factors such as donor organ age and agonal warm ischemia may become additive insults, on top of preservation injury, which can negatively influence the long-term outcome of the graft.

At present, kidneys are the most frequently used organs from non-heart-beating donors, although liver donation is also possible in selected cases. In relation to organ preservation, it is usual practice now for kidneys from non-heart-beating donors to be assessed during hypothermic perfusion preservation. This allows the possibility of avoiding the transplantation of organs that have been seriously damaged by the warm ischaemic insult, on the grounds of poor perfusion characteristics or high levels of released enzymes (indicating renal cell damage) into the perfusate. However, there are not yet universally accepted viability tests on which to base this decision. Perfusion preservation also allows the possibility of the resuscitation of the prior ischaemic damage in organs; for example, by stimulating resynthesis of ATP via aerobic metabolism, which is possible even at hypothermia. This has been demonstrated in many experimental studies but has yet to be established in the clinical situation, although the value of hypothermic perfusion

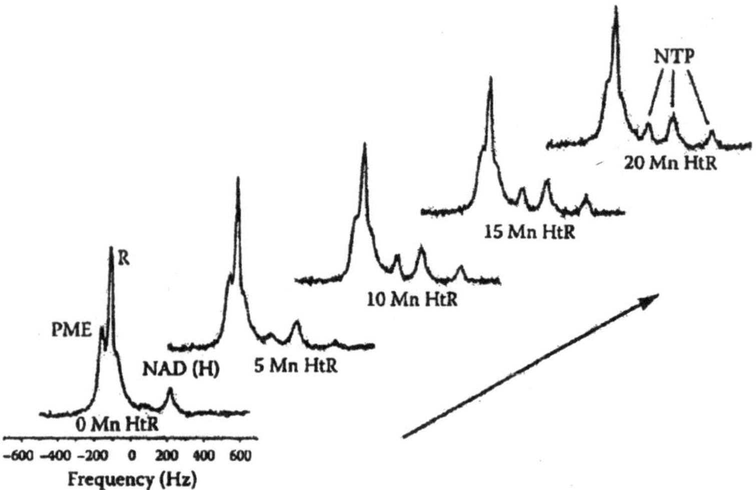

Fig. 8.5. Typical ^{31}P-MRS spectra of a flush-harvested pig liver during the period of hypoxic cold storage and hypothermic reperfusion (HtR).

versus simple cold storage has been shown in some clinical reports. Predictions of improved long-term results using cold perfusion have long appeared in statistical models, but as this is a relatively recent activity in transplantation, the true results will not be available for some years.

Steatosis

This is a complication that mainly affects liver preservation. The term steatosis (or fatty liver) identifies a liver in which lipid, mainly triglyceride, accounts for more than 5% of liver wet weight. Hepatic steatosis may develop without clinical or biochemical evidence of liver disease. Biochemical plasma liver function test results frequently fail to identify a liver with even a considerable degree of steatosis. However, after cold preservation, such a liver is at substantial risk for primary non-function, or at least a period of very poor function after reperfusion. With the shortage of donor organs, steatotic livers are used increasingly for transplantation. The rate of steatosis found in potential living-related liver donors, who represent a relatively healthy nonbiased population, is 6–10%. Far greater rates of fatty change were found in studies dealing with victims of trauma deaths, reaching 24% in victims of fatal traffic accidents and 30% in children who had died in traumatic accidents who had no preexisting comorbidity. These very high levels of steatosis may reflect relative hypoperfusion of the liver in a shocked patient after trauma.

The mechanisms behind this rate of change are still not completely understood. Experimental studies of steatotic animal models showed an inverse correlation between degree of steatosis and sinusoidal blood flow, believed to be caused by ballooned hepatocytes containing fat droplets that compress and distort the sinusoidal lumen and increase intrahepatic portal resistance. This resistance in turn leads to relative ischemia of fatty hepatocytes, shown to have increased sensitivity to anoxia in culture, and thus exaggerating the susceptibility of these organs to preservation–reperfusion injury. This shunting of blood from the hepatic microcirculation may be responsible for impaired perfusion of an organ with cold preservation solution, leading to more damage during the preservation process. Inefficient anaerobic metabolism during warm ischemia and cold preservation, with the depletion of high-energy phosphates and the buildup of lactic acid, is believed to be one of the major mechanisms contributing to reperfusion injury.

The energy level within the hepatocyte after the period of preservation has been shown to correlate with the eventual outcome after transplantation. Several investigators have shown abnormal adenosine triphosphate production within the mitochondria of steatotic livers. This is believed to be caused by the accumulation of nonesterified fatty acids and has been postulated as a possible mechanism for the increased sensitivity of fatty livers to ischemic injury. Others postulate that the critical injury is to the sinusoidal lining cell, with an alteration in plasma membrane fluidity and subsequent leukocyte adhesion and Kupffer cell activation. Others suggest that fat-laden hepatocytes undergo changes, including solidification during cold injury, and are responsible for increased sensitivity to reperfusion injury, releasing fatty globules that can disrupt the sinusoid microcirculation. The massive accumulation of lipids released from hepatocytes within sinusoids after transplantation, mimicking peliosis, has been termed lipopeliosis.

Cellular disruption, with free-radical formation and triglyceride and free–fatty acid release, is known to activate phospholipase digestion and lipid peroxidation, thus causing additional cellular damage. It also has been shown in an experimental animal model that steatotic livers are more susceptible to injury during subsequent periods of warm ischemia. Disruption of the sinusoidal microvasculature again is postulated to be the underlying mechanism. In view of the increased sensitivity of steatotic livers to both warm and cold ischemia, the logistic difficulty of transplantation of this group of organs increases, with both ischemic times required to be as short as possible, thereby minimizing the risk of primary non-function. These constraints, together with recipient factors that may prejudice outcome, need to be taken into account when a decision is made to use a steatotic organ.

Hepatocyte growth factor has ameliorated preservation injury in fatty livers, possibly by providing some protection to hepatocytes and sinusoidal lining cells. Prostaglandin E_1, an agent known to improve hepatic microcirculation, also has shown a beneficial effect in diminishing ischemic injury in a Zucker rat fatty-liver transplant model. The use of oxygenated perfusion to predict survival of steatotic livers before transplantation would require establishing parameters that correlate with survival.

These may possibly include hemodynamic parameters as well as markers of synthetic function and liver injury assessed by sampling perfusate. If a supply of oxygen and nutrients is provided within a suitable perfusate, then continual normothermic preservation may become a viable reality, thus abrogating the increased sensitivity of fatty livers to cold-preservation reperfusion injury.

Development of Preservation Solutions

The development of preservation solutions for clinical use has resulted from a combination of experiments in tissue biochemistry and experimental transplantation. The work of Keller et al. demonstrated that ion diffusion and edema in renal tissues could be usefully investigated during cold preservation, and these techniques were used by many later authors, resulting in several different preservation solutions being reported over the last 30 years. In particular, prolonged preservation became possible after the development of solutions of intracellular electrolyte composition. These solutions minimized cellular edema and the loss of intracellular potassium.

Collins ushered in a new era of simple hypothermic storage by showing that a solution of crystalloid solutes alone, more resembling intracellular than extracellular composition, and quite free of colloid, could extend renal preservation reliably for 48 hours and longer. Subsequently, solutions of widely different composition, based on the citrate anion, amino acids (Bretschneider), or simple sugar (sucrose), were shown to be equally or more effective than the original Collins' solution, and the significance of intracellular composition became less relevant. The composition of equally effective flush solutions can differ significantly, but each solution has components aimed at preventing cellular swelling, and each contains a buffer. These solutions are used primarily in static hypothermic storage. Various adjuvant solutes can further enhance preservation.

Static Hypothermic Storage: Intracellular Solutions

Collins' and EC solution

Many preservation solutions were developed along with research efforts to find appropriate car dioplegic solutions, and one of the first solutions for kidney preservation was developed by G.M. Collins in laboratories in Australia and Los Angeles, California. Prolonged kidney preservation by simple cold storage became possible after the development of Collins' intracellular solutions. However, after 72-hour cold storage, the initial function of kidneys was poor, and any added warm ischemia markedly diminished the efficacy of these solutions. The early Collins' solutions (C2, C3, and C4) had high concentrations of potassium, magnesium, phosphate, sulphate, and glucose (120 mM). Precipitation of magnesium phosphate was a major technical defect. Additions of procaine (in C3 and C4) and of phenoxybenzamine (C4) were found to be unnecessary and, in the case of procaine, harmful. Subsequently, mannitol, magnesium, procaine, and phenoxybenzamine were omitted from the very widely used EC solution, containing high concentrations of potassium and phosphate and increased glucose. The higher glucose concentrations reduced the potential efficacy of EC solution and made it less effective in the preservation of organs other than the kidney.

Glucose is slowly permeable across cell membranes, and even under hypothermic conditions, glucose is broken down at least to the level of lactate, which augments tissue acidosis and doubles the intracellular carbon solutes, promoting cell swelling on an osmotic basis. These potentially harmful effects become more apparent with prolonged storage times. Replacement of glucose by other less permeable and nonmetabolizable solutes (sucrose, mannitol) improved the results of renal preservation by Collins' solution in rat and dog kidneys. A significant reduction in post-transplantation renal failure occurred when human kidneys were preserved in EC solution with mannitol substituted for the glucose.

The replacement of glucose by sucrose may be even more relevant in liver and pancreas grafts, in which cells are more permeable to glucose. EC solution, although cheaper, was not simple to use, because sterilization of the glucose-containing fluid had to be done separately from the electrolyte part and mixed before use. EC solution was standard in Europe for about 15 years.

Citrate-based solutions

Citrate solution was developed to overcome some of the limitations of Collins'-type solutions. This solution also has high concentrations of potassium and magnesium, but citrate replaces phosphate and mannitol replaces glucose. Citrate provides buffering capacity and interacts with magnesium to form impermeable, stable chelates. A high concentration of magnesium is necessary for the efficacy of citrate solutions; only 4% of the magnesium remains unchelated, and magnesium-citrate chelate contributes to the semipermeable solute available physically to maintain cellular integrity.

High levels of potassium may minimize the loss of cellular potassium. Citrate chelates calcium as well as magnesium and may aid in the extrusion of calcium from the cell as well as the maintenance of pH; citrate could be replaced by its analog, tricarballylate, provided hydroxyethylpiperazine-ethane sulphonic acid (HEPES) is added to augment its buffering capacity. Isotonic and hypertonic solutions are equally effective in preserving human kidneys and livers; the sulphate ion is unimportant and can be replaced by other anions. Mannitol can be replaced by sucrose with equivalent results; the replacement of mannitol with glucose was detrimental in kidney preservation.

Effective buffering of the solutions is a most important feature. An acid pH (6.8) is deleterious, whereas solutions with a pH of 7.1–7.8 were equally effective. Citrate solutions gave results comparable to Collins' solutions experimentally and clinically for kidney, liver, and pancreas grafts.

Sucrose solutions (Phosphate-buffered sucrose)

The demonstration that sucrose improved efficacy of preserving solutions led to the formulation of a simple isosmolar sodium phosphate-buffered sucrose solution (PBS). This very simple solution provided excellent preservation of kidneys. Sucrose-based solutions, optimally at a sucrose concentration of around 140 mM, are also effective in preservation of other organs. These solutions contain no potassium or magnesium. Addition of gluconate does not improve the efficacy of the solution.

Hyperosmolar sucrose-based solutions are less effective than isosmolar solutions. These sucrose solutions provide buffering and impermeable solute only; cooling reduces metabolic requirements, and no other physicochemical attributes can be demonstrated. This indicates that hypothermia, buffering, and impermeable solute represent major components of static preservation in kidneys.

UW solution

Research by Folkert Belzer and James Southard, acknowledged pioneers of organ preservation, resulted in a development of a new preservation solution called UW, initially developed for the pancreas, that introduces three new philosophies: osmotic concentration is no longer maintained by metabolically active glucose but, rather, is achieved by the administration of metabolically inert substrates like lactobionate and raffinose; much attention is paid to the additional administration of colloid hydroxyethylstarch (HES); and addition of oxygen radical scavengers, glutathione, allopurinol, and adenosine in addition to steroids and insulin

Subsequently, UW was effectively applied to liver, kidney, and heart preservation. The clinical success of UW in liver, pancreas, and cluster preservation has made it the preferred solution for multiple-organ harvest. At present, UW is considered to be the most generally effective flush solution. It was the first to contain lactobionate and raffinose, both relatively large impermeant molecules that effectively control cell volume. Lactobionate seems to be a rather unique component, the full action of which remains to be explained, possibly it is based on the chelation of calcium or iron. UW and Collins' solutions both contain phosphate to control acidosis and a high potassium level to minimize loss of intracellular cation. However, these features may be of relatively minor importance because the small volume of solution left in an organ probably has little buffering capacity in comparison with the much larger amount of intracellular buffer, and solutions with an extracellular cation content have been found to work well for liver, kidney, and pancreas preservation. The requirement for colloid in flush solutions is controversial. It may prevent interstitial edema, but at the price of appreciably higher solution viscosity. The use of a colloid has been particularly advocated for pancreas preservation.

Although the current supremacy of liver storage by UW has become well established, and assessments of the advantages of this solution have been recorded on clinical and histological grounds, a complete explanation for its efficacy remains obscure. One mechanism may lie in synergism between various UW ingredients. For example, adenosine and glutathione exert greater hepatocyte protection when used together than when either is used alone, and other such links between UW components may provide a summation of protection. However, removal of some UW ingredients does not affect the quality of cytoprotection, and substitution of other ingredients with substances postulated to have a similar effect does significantly alter preservation.

Better heart, lung, and kidney preservation by UW, compared with EC solution, is much less dramatic than improved liver preservation. This indicates either a mechanism of hepatic cytoprotection by UW that is distinct from the prevention of hypothermic damage incurred by all cells, or a liver-specific susceptibility to hypothermia that is selectively retarded by UW support. Cold storage indeed produces a specific injury to the hepatic sinusoids. Morphological studies of rat grafts after periods of cold storage and reperfusion show that hepatocytes are very largely preserved and Kupffer cells are activated, and the dramatic, predominant histological finding is gross destruction of the hepatic sinusoidal endothelial lining.

Endothelial cell death is delayed after storage in UW compared with storage in EC solution. Furthermore, videomicroscopy of cell detachment, mitosis, and motility confirms better preservation of endothelial cell stability after UW than after EC preservation. The liver exhibits highly specialized microcirculatory features. For instance, distinctive endothelial fenestral actin rings may contract and relax to alter vascular and Hisse space perfusion. Specialized unfenestrated endothelium at the narrow junction of the sinusoidal space and terminal portal venule may protrude into the vascular lumen to occlude it. Such an inlet sphincter may be enhanced by a local periendothelial contraction of Ito cells. The highly specialized feature of liver sinusoidal endothelium, and its particular susceptibility to hypothermic injury, might account for this selective improvement of liver preservation by UW and the less marked improvement in UW preservation of other organs.

UW incorporates a variety of additives: adenosine (ATP precursor and vasodilator), allopurinol (xanthine oxidase inhibitor), glutathione (free-radical scavenger), insulin, and steroid (membrane stabilizer). The solution has a high concentration of potassium. The relative importance of individual components has been elucidated in a number of studies. Modifications of the original solution with the omission of one or more components have provided equivalent preservation of kidney and other organs. No modification has been shown to improve preservation significantly, but simplification of the base solution has allowed additional room for future additives that may prove helpful. It is worth considering the important components of UW and their relative importance.

HES

The inclusion of the synthetic colloid HES does not appear to confer any advantage. Colloid added to flushing solutions has long been known to improve the rheological properties and the efficiency of vascular washout, but the addition of colloid is only essential in continuous perfusion storage. Colloid increases the viscosity of the solution, but this does not appear to be either harmful or helpful in simple flushing. Preservation of canine kidneys by simple hypothermic storage was reliably extended to 5 days by a modified HES-free UW. Omission of colloid in a randomized clinical trial showed equivalent and satisfactory preservation of kidneys with HES-free UW. Both UW and its colloid-free modification gave significantly better preservation than EC.

Electrolytes

Reduction of potassium concentration with reversal of the sodium/potassium ratio considerably improves flushing efficacy (potassium is markedly vaso-constrictive) and is just as effective in preservation. A modification without HES and based on sodium lactobionate and sucrose with low potassium has been widely used experimentally and clinically for multiple-organ preservation.

Impermeants

The great efficacy of UW in organ preservation is primarily a result of the impermeable anion lactobionate. Gluconate, also impermeable, can replace lactobionate. Similarly, the impermeable raffinose can be replaced by sucrose with little detriment.

Other additives

The omission of insulin and dexamethasone has not been detrimental. Adenosine, allopurinol, and glutathione were found to confer some benefit in the preservation of rat kidneys, but both adenosine and allopurinol were omitted in a clinical trial with no detrimental effects.

The introduction of UW and its modifications has markedly enhanced preservation of all transplantable organs; it has restored order to the operations of liver and pancreas grafting, given more effective heart preservation than previous methods, and served as the standard for all transplantable organs. Although differences between organs in their sensitivity to ischemic damage and their individual requirements for optimal preservation certainly exist, these differences are out-weighed by the universally protective effects of the major components of UW.

UW has proven to guarantee extreme long cold-ischemia tolerance — 72 hours in kidneys and certainly 48 hours in livers. This solution was also tested prospectively in a multicenter European trial, in which UW was found to be superior to EC by decreasing the rate of delayed graft function from 33% to 23%. The clinical success of UW and its variants in heart and lung, liver, pancreas, kidney, small bowel, and organ cluster grafts has made it the preferred solution and the gold standard for multiple-organ harvest in the 1990s.

A multicenter trial in Europe comparing UW with EC showed that UW preservation gave more rapid recovery of renal function and less posttransplant dialysis; other studies showed equivalent function of kidneys preserved by UW and Collins' solution in a randomized prospective clinical trial. UW has been shown to diminish the trapping of erythrocytes in the microcirculation. The effects of UW on protecting nonparenchymal cells in kidney, liver, and pancreas may be important aspects of the solution's abilities to provide extended preservation, rather than effects on the more easily preserved parenchymal cells.

Bretschneider's histidine-tryptophane-ketoglutarate solution

Histidine-tryptophane-ketoglutarate solution (HTK) was introduced by H.J. Bretschneider from Gottingen, Germany, primarily as a physiologic cardioplegic solution, and was subsequently shown to be effective in the experimental and clinical preservation of the kidney and liver. HTK contains relatively impermeable amino acids and other solutes; its main component is histidine, with added mannitol, tryptophane, and alpha-ketoglutarate, with low electrolyte concentrations of sodium, potassium, and magnesium. Histidine is an excellent buffer, and tryptophane, histidine, and mannitol can act as free-radical scavengers.

The HTK solution has a very low viscosity, and according to Bretschneider, high volumes at a low flow rate should be applied to guarantee "*equilibration.*" HTK has been tested in clinical trials. The solution has given a low incidence of delayed function in kidney transplantation clinically and was shown to be better than EC in kidney transplantation in a prospective randomized clinical trial. Eurotransplant conducted a multicenter randomized prospective trial comparing UW with HTK; the results show that both solutions have the same efficacy at the endpoint of delayed graft function. Clinical use of HTK in liver preservation is documented by single-center experiences, which show the same safety and efficacy profile as UW, at least in the time range of below 24 hours cold ischemia time.

Static Hypothermic Storage: Extracellular Solutions

Celsior

The Celsior preservation solution created by Pasteur-Merieux is a mixture of the impermeant inert osmotic carrier philosophy from the Belzer solution, using lactobionate and mannitol, and the strong buffer philosophy from the HTK solution, using 30 mmol of histidine buffer. In addition, Celsior also contains oxygen radical scavengers. From the electrolyte point of view, Celsior is a high-sodium, low-potassium solution that is unique in this respect. Celsior has been used clinically in cardiac conservation. Interestingly, the heart seems to require a high- potassium solution, which raises the question as to whether there may be other organs that have specific requirements.

Other extracellular solutions

Solutions of extracellular electrolyte composition, such as plasma, saline, or Hartmann's solution, are unsuitable for prolonged (>12 hours) preservation but are often used to flush-cool kidneys *in situ* during multiple organ procurement. They may also be used for cooling kidneys before immediate transfer in living donor transplantation but are not recommended because alternative solutions have been demonstrated to be markedly superior. Wallwork and low-potassium dextran solutions are used in lung preservation as a variant of UW that may be considered an extracellular solution; namely, reverse Na^+/K^+ UW or low-potassium UW. Stanford, Hopkins, and St. Thomas Hospital solutions are extracellular solutions used in heart preservation.

Summary of Preservation Solution Development

In summary, EC has almost been abandoned because of the glucose disadvantage, whereas UW is certainly the most used preservation solution for livers, kidneys, and pancreases, with excellent clinical and experimental preservation data. UW can certainly be considered the current gold- standard solution. However, it is not without its disadvantages — the high viscosity, price, and difficult handling of numerous 1-L bags, and the fact that the radical scavenger glutathione is gradually oxidized in the bags (presumably because of diffusion), encourage competitors to produce new compounds with better cost-to-effect ratios. HTK has a firm place in cardiac preservation by demonstration of equal safety and efficacy in preserving livers and kidneys, at least in the middle and lower ranges of cold-ischemia time. HTK will be used more frequently, particularly with the consideration of lower price and more easy handling aspects. The suggested high-volume perfusion is not really necessary, and a calculation based on a total volume of 10 L for a multiorgan donor shows significant cost reductions. Celsior is currently only used for cardiac preservation.

Finally, although the basic aim of improved solutions is to extend storage times, clinical application of this stratagem may be unwise because there is a relationship between storage time on the one hand and the incidence of delayed function, rejection, and poorer long-term results on the other. In an analysis of the influence of the ice storage time of first cadaveric renal transplants, Collins' suggested that the best results were seen with the shortest preservation times. In this setting, living donor transplants do no better than what is projected for cadaveric transplants if storage times could be reduced to only 3 hours. The inference is that the immunologic significance of the preservation effect is of greater importance than histocompatibility, and it may be worthwhile to search for components such as polyethylene glycol that affect long-term outcome as well as immediate function.

Machine Perfusion Preservation

The development of *ex vivo* perfusion as a means of sustaining organ function was extensively evaluated by the Nobel prize-winning scientist Alexis Carrel and the aviator technologist Charles A. Lindbergh in the 1940s. Their aim was to simulate normal physiologic conditions with *ex vivo* perfusion techniques. This concept was modified by Ackermann and Barnard, who provided the isolated organs with continuous low-flow arterial circulation, using a perfusate primed with blood and oxygenated within a hyperbaric oxygen chamber.

This technique was initially most widely used during the early days of clinical renal transplantation. It has remained the preservation method of choice in some centers, and there is now a resurgence of interest in the storage and assessment of ischemically damaged kidneys obtained from non-heart-beating donors. Continuous perfusion aims to simulate conditions *in vivo*: The organ can be continually supplied with oxygen and nutrients; waste products are continuously removed. However, the method is complex, costly, and more liable to complication (by technical error or aseptic breach) than is simple static cold storage. Perfusion techniques still provide the maximum extension of successful storage of most organs. The essential requirements are sterile nonthrombogenic and nonallergenic delivery tubing, an organ chamber, a pump and heat exchanger, a mechanism for oxygenation (membrane oxygenation or surface diffusion), appropriate

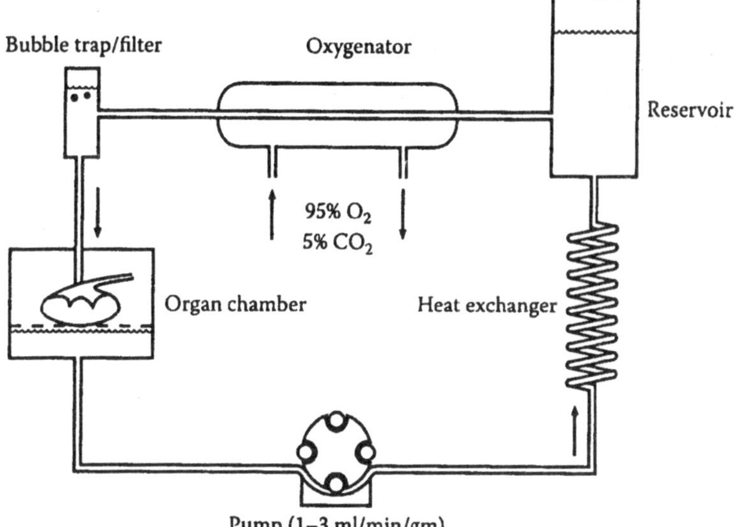

Fig. 8.6. A schematic representation of a generic organ perfusion circuit.

monitoring of perfusate temperatures and pressures, and an effective perfusion fluid.

However, the complexity of the method precluded its general use. Elimination of the hemoglobin and hyperbaric chamber components by Belzer et al. resulted in satisfactory kidney preservation for up to 2–3 days. The asanguineous perfusion technique eventually was abandoned in most kidney transplant centers when it was learned that the quality of 2-day preservation was no better than with the simpler static methods. Nevertheless, it is expected that the refinement of perfusion technology will someday permit true organ banking.

Stable solutions were needed to enhance the rapidity and uniformity of cooling by perfusion and to provide solutes that improved preservation of vasculature, parenchyma, and interstitium. Blood was demonstrably not the ideal medium: Cold blood adds rheological problems of greater viscosity and sluggish flow. Simple balanced electrolyte solutions mimicking the extracellular milieu, such as Ringer's-lactate solution or saline, were shown to be unsatisfactory organ-preserving solutions that exacerbated interstitial and cellular edema. For continuous perfusion, attention was then directed by Belzer and others toward plasma-like solutions containing albumin, dextrans, starch, or other colloids (combined with extracellular electrolyte content) to provide oncotic support and to prevent interstitial edema.

Cold perfusion has shown a greater capacity to revive and resuscitate organs previously damaged by warm ischemia. This is this method's area of greatest potential, together with the prospect of potentially infusing immunosuppressants or agents capable of deleting antigen-presenting cells before implantation.

All perfusion storage is carried out between 4° and 9°C, and the perfusion circuit must include a heat exchanger to maintain hypothermia. A reliable pump is essential, and the original Belzer pump provided physiological pulsatile flow, but such flow does not appear to be essential. Nonpulsatile pumps have been equally effective. The commonly used equipment provides nonphysiological pulsatile flow at a pressure of 20–60 mm Hg and an average flow of 1–6 mL/g per minute.

Oxygenation is a feature of some perfusion storage circuits. Some systems include a membrane oxygenator (e.g., Belzer and early Waters machine); other systems rely on surface oxygenation (e.g., Gambro) and appear to be equally reliable. A mixture of air or oxygen with added carbon dioxide is essential for membrane oxygenators to maintain pH as well as oxygen tension. Surface oxygenation is possible without added carbon dioxide. The perfusates must pass through a bubble trap before reaching the organ. More recently, air equilibration without

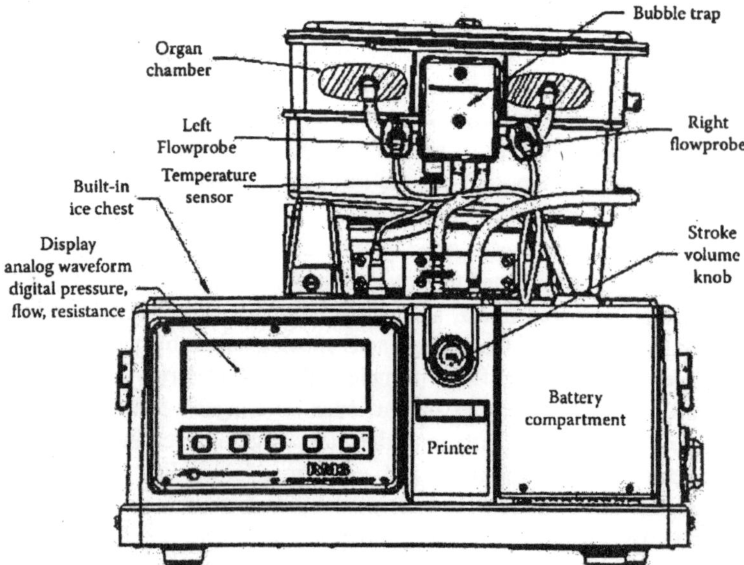

Fig. 8.7. A schematic representation of a Waters kidney perfusion circuit.

added carbon dioxide has become routine, with the introduction of Belzer's machine perfusion solution, and a new generation of portable hypothermic perfusion machines has been developed.

Blood has no advantages and considerable disadvantages in hypothermic perfusion. Cold blood introduces problems of increased viscosity, resistance to flow, and aggregation of platelets and red blood cells.

Cryoprecipitated plasma derived from blood treated with the anticoagulant acid-citrate-glucose, with added magnesium sulphate and mannitol, provides good preservation for 3 days. However, the preparation of this perfusate is time consuming and difficult to standardize; thus, other plasma-derived solutions were developed — plasma protein fraction and silica-gel fraction plasma — but these have largely been superseded by the gluconate-based (Belzer-type) solutions.

Colloids

A colloid of molecular size similar to or exceeding that of plasma protein is essential for the continuous machine perfusion preservation needed for providing oncotic vascular support. Omission of colloid leads to progressive transcapillary passage of perfusate into the interstitium by ultrafiltration, causing explosive weight gain and edema of the organ, increased resistance to perfusion, and early organ failure.

Machine perfusion has been applied most extensively to kidney preservation. Belzer and colleagues in the 1960s demonstrated the effectiveness of cryoprecipitated plasma thawed and passed through a membrane filter. Subsequently, perfusates based on purified albumin (MW 68 kDa) were shown by Johnson and Toledo-Pereyra to be equally effective.

Substitutes for these biological colloids were sought to make perfusate preparation easier and to eliminate the risk of transmitted viral infection. The most widely applied synthetic colloid, subsequently used by Belzer and Southard in developing UW, was HES. A UW developed for continuous perfusion, containing HES, has been effective in extending the preservation of canine kidneys by machine perfusion for 7 days. Dextrans are synthetic macromolecules derived from bacterial digestion of starch, and they occupy a spectrum of molecular size from 40 to 150 kDa. Perfusates using Dextran 70,000 or Dextran 40,000 have been successful in the preservation of kidneys and other vascularized organs. PEG, an impermeable polymer also available in a range of molecular sizes, has been used by Collins to develop solutions suitable either for perfusion or for flushing. PEG also has been suggested to have additional effects in diminishing the intensity of the immune rejection response. Colloids, although essential for perfusion preservation, are neither necessary nor advantageous for simple hypothermic storage. In any case, the colloid in a flush solution is usually regarded as an inert component included only for its oncotic properties. PEG appears to be an exception to this rule. When PEG 20,000 was substituted for HES in UW, the resulting solution, known as Cardiosol, was found to be significantly superior to UW for 24-hour heart preservation.

Impermeable Solutes

Solutes that are impermeable or are only slowly permeable to cellular membranes are a vital component of preserving solutions — whether used for cold-flushing or for continuous perfusion. Impermeable solutes are required at concentrations of up to 140 mM to counteract the effect of diminished cell membrane ion transport in the cold. Large inert molecules, metabolically and nutritionally unavailable to cellular physiological processes, have been most effective. Gluconate (for perfusing and flushing solutions, MW 196) and lactobionate (for flushing solutions, MW 358) have been very successful; other polysaccharides and disaccharides (raffinose, MW 594; sucrose, MW 134; maltose, MW 360; trehalose, MW 378) and chelates of citrate and magnesium have also been very effective for static flush but not so important for continuous perfusion, and metabolizable monosaccharides (glucose, fructose) have been less effective, and colloids alone are not effective.

Other agents, which diffuse into cells, but at a slower rate than do simple electrolytes, have also been used. Amino acids (glutamine, MW 146; histidine, MW 155; tryptophane, MW 204; aspartate, MW 133; glycine, MW 75) and fatty acids (linoleic acid, MW 280) have also been used in successful solutions.

Electrolytes

Originally based on extracellular solutions that were high in sodium and chloride, the composition of preservation solutions subsequently saw a swing of the pendulum to solutions that were high in potassium and phosphate (intracellular type) and then back toward normal extracellular concentrations of sodium and potassium as some of the potentially deleterious consequences of very high potassium levels were appreciated. Cold-induced paralysis of cell membrane pumps requires the chloride anion to be reduced and replaced by less permeable anions — gluconate, lactobionate, citrate, and phosphate.

Apart from these measures, electrolyte content is of relatively little importance. Potassium in the original intracellular concentrations of 125–135 mM is markedly vasoconstrictive and potentially harmful to the vascular endothelium. Solutions lower in potassium and higher in sodium (sodium 130 mM, potassium 5–30 mM) have been equally effective. Magnesium has been a useful additive in some solutions. Calcium tends to precipitate with phosphate and has been excluded from solutions or added in minute concentration. Successful solutions have been isosmolar or hyperosmolar.

The hydrogen ion concentration again has not been a critical factor, although extremes of acidity and alkalinity have diminished effectiveness. All successful solutions use one or more buffers — bicarbonate, phosphate, citrate, and many others — but intracellular acidosis and its correction have not wholly correlated with effectiveness of solutions.

Oxygenation

During perfusion storage, some nutrient substrate has been regarded as essential, together with continuous oxygenation. Oxygen can be supplied by the use of membrane oxygenators or by passing a stream of humidified gas across a thin film of flowing perfusate. At hypothermic temperatures, equilibration with air may provide sufficient oxygen to stimulate aerobic energy production in some situations, but for large organs, extra oxygenation may be essential. Mixtures of oxygen and carbon dioxide can also be used to stabilize the pH of the perfusing solution if a bicarbonate buffer is included, but newer perfusion solutions, such as the Belzer gluconate-based solution, do not need carbon dioxide for pH control.

The addition of oxygen during storage can also be achieved by another technique, which is not classic perfusion, but does use the vascular bed of the organ — this technique is called venous oxygen persufflation. In venous oxygen persufflation, cold, humidified, oxygen-rich gas is supplied at hypothermic temperatures and low controlled pressure via the venous system (e.g., in the kidney, via the renal vein). Fine-needle punctures on the surface of the organ allow a more even diffusion of gas to peripheral regions and the escape of gas during storage. This method is sufficient to promote ATP synthesis. The organ may need to be immersed in preservation solution to avoid drying. Persufflation of the kidney during storage has shown some experimental advantages, especially in organs that have been damaged by prior warm ischemia, but only one clinical trial has so far been undertaken.

Another means of enhancing oxygen supply is the use of synthetic perfluorochemical polyols to substitute for erythrocytes in facilitating oxygen transport. These agents have been used for continuous machine perfusion and also for static storage, in which oxygenation by diffusion may be achieved in smaller organ masses such as the pancreas. In this latter procedure, these agents have been helpful in enhancing oxygen delivery to stored organs balanced at the interface between two solutions of different physical characteristics — one containing the synthetic fluorochemical. The method has been shown to enhance pancreatic storage in particular.

The application of oxygenation during cold preservation, particularly by persufflation, does raise the issue of benefit (sustained aerobic energy metabolism) versus risk (oxidative stress). To this end, it has been shown that the addition of extra anti-free-radical agents (in addition to glutathione present in many preservation solutions) can have beneficial effects. For example, superoxide dismutase has been shown to improve experimental oxygen persufflation.

For perfusion, substrates such as glucose (10 mM) are often used, sometimes enriched with fructose, pyruvate, amino acids, fatty acids, phosphates, adenosine, and other high-energy compounds or precursors; for example, the addition of adenosine- and phosphate-enhanced ATP synthesis during hypothermic liver perfusion. However, even during optimal oxygenated hypothermic machine perfusion, aerobic metabolism may not be fully restored during storage.

Additives to Preservation Solutions

In general, colloid is necessary for machine perfusion, but not for simple storage. Extended preservation times by perfusion require suitable nutritional precursors and oxygen supply. Otherwise, prerequisites for machine perfusion solutions, and for solutions used for cold flushing before hypothermic storage, have been similar. The major requirements have been suitable impermeable solutes (anions or non-electrolytes) replacing permeable chloride, together with a suitable buffer system.

The role of nutrient substrates during static storage, and on reperfusion, is also unclear. Substrates are not obligatory — solutions containing no utilizable metabolic fuel can give excellent preservation of most organs for 24 hours, and in the case of the kidneys, for up to 3 days experimentally with PBS solution. In contrast, substrates capable of supporting anaerobic glycolysis (such as fructose) have been shown to be beneficial in some experimental situations.

Static storage normally implies anaerobic conditions with progressive fuel depletion. Precursors of ATP added to flushing solutions (such as adenosine, adenine, inosine, and phosphate), ATP itself, or high-energy intermediates such as fructose 1,6 bis phosphate have been shown helpful in some experiments. Whether these agents are acting by increasing the availability of high-energy compounds or by some other metabolic or hemodynamic manner is uncertain. In contrast, nutritional depletion in experimental conditions can enhance tolerance to ischaemic damage. Glycogen-depleted livers from fasted rats showed resistance to ischemic injury and better survival after transplantation.

Small concentrations of added agents that quench or scavenge oxygen free radicals, antioxidants, calcium-channel blockers, and other pharmacologically or metabolically active compounds can help as adjuvants in increasing ischemic tolerance. In UWs, allopurinol, glutathione and adenosine have been demonstrated to be helpful — insulin and steroids appear of no benefit. Other additives have included cytokines, prostaglandins, leukotrienes, lazaroids (aminosteroids) and their intermediaries, hormones, chlorpromazine and other benzodiazepines, nitric oxide, interleukins, TNF antagonists, and other growth factors. For example, prostaglandin analogue (Iloprost) was shown to enhance ATP recovery when added to the cold reperfusion

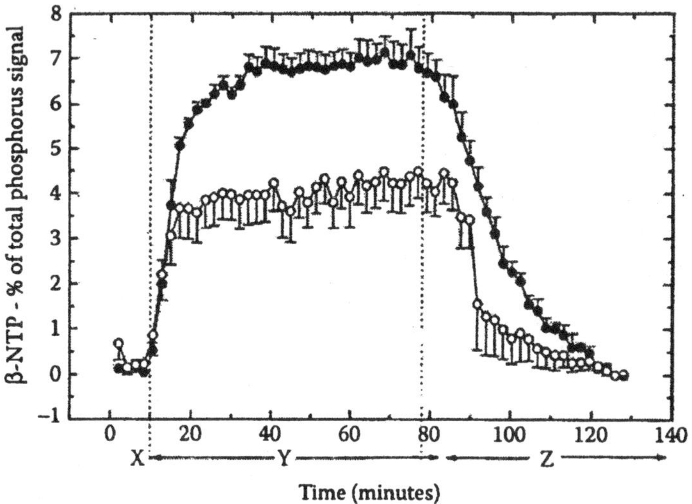

Fig. 8.8. Changes within ATP following hypothermic reperfusion with Iloport.

solution in experimental porcine liver preservation. Most recently, the addition of a mixture of trophic factors during cold preservation has shown promise in experimental work.

Storage-reperfusion injury affects endothelial cells and activates neutrophils and macrophages (e.g., Kupffer cells in liver). This injury leads to the expression of cell adhesion molecules, leukocyte and platelet adhesion, and exacerbation of reperfusion injury. Kupffer cells are a main source of oxygen radicals and also release other proinflammatory mediators, such as TNF-α, interleukins 1 and 6, eicosanoids, and nitric oxide. Agents suppressing TNF-α formation include calcium channel blockers, pentoxifylline, adenosine, and prostaglandin E, whereas liposaccharides from bacteria are stimulatory.

Storage Temperature

Organs surrounded by preserving solution and stored and transported in a container of crushed ice maintain a storage temperature of melting ice (0°–2°C). Storage temperatures above 0°C aim to diminish cold-induced membrane paralysis while maintaining the protective effects of hypothermia. Temperatures above 15°–20°C have usually been considerably less effective than 0°C. Temperatures for continuous perfusion have been maintained between 4°C and 10°C. The optimal storage temperature of nonparenchymal cells may differ from that required by parenchymal cells, perhaps over quite a narrow range. In rat liver preservation, 48-hour storage was possible by simple hypothermic storage at 4°C but not at 0°C. Differences in effectiveness between various solutions (UW, sucrose, citrate) were less marked during storage at 4°C than at 0°C.

These results indicate a greater sensitivity of endothelial cells to cold damage than occurs with parenchymal cells, and temperatures around 4°C may be optimal for hypothermic storage. Storage at 4°C could be simply obtained by the portable electrical refrigerator storage of organs surrounded by cold liquid rather than by ice, but this would require good temperature control within the system, as air is a poor conductor of heat. Such equipment has been used in experimental studies but not clinically examined yet to any great degree.

Pre-reperfusion Rinse Solutions

A prevascularization flush-out or rinse at the conclusion of the storage period, just before reimplantation, has two main aims: first, to remove potentially toxic components of the preservation solution and accumulated waste products (e.g., high levels of potassium and hydrogen ions) or to prevent a large intracardiac bolus of cold fluid on clamp release. Second, rinsing aims to smooth the transition between hypothermic storage and isothermic revascularization. Anastomosis can be tested before revascularization by intravascular wash-out before final clamp release. This process has been most studied in relation to the liver. One method allows blood to flow to the liver via the portal vein, venting the initial effluent from the infrahepatic vena cava before completing this anastomosis, and finally releasing the clamp on the suprahepatic vena cava. Alternatively, the organ can be reflushed via the portal vein or hepatic artery with a cold or warm electrolyte solution (e.g., Ringer's-lactate) or with albumin before clamp release.

Carolina rinse solution (CR), developed by Lemasters and colleagues at the University of North Carolina, maintained the basic electrolyte composition of Ringer's lactate, but had added components and provided better results than a simple electrolyte rinse. CR combines extracellular electrolyte composition with a mildly acidic pH, oncotic support (albumin or HES), antioxidants (allopurinol, glutathione, desferrioxamine), energy substrates (adenosine, glucose and insulin, fructose), vasodilators (calcium channel blockers), and glycine. As with UW, it was not clear which of the several components were effective and whether the effects were the result of the beneficial effects of the CR or the potentially harmful effects of a Ringer's-lactate rinse.

Analysis of the various components and techniques indicates that the temperature of the rinse solution is preferably above the required cold-storage temperatures of 0°–4°C; rinses at ambient temperatures (20°C) and at body temperatures (37°C) have given better results. The preferred pH of the rinse may be slightly acidic (pH 6.8). Components that seem unnecessary are colloid (HES or albumin), nicardipine and magnesium, glucose, fructose, and insulin. Beneficial agents include adenosine, allopurinol, desferrioxamine, glutathione, and glycine, possibly acting through free- radical scavenging and quenching.

Similarities between these findings and those with UW are apparent. Whether a CR rinse has advantages over a prereperfusion second rinse with a UW-derived, low-potassium, sodium lactobionate and sucrose solution used for storage is not yet established, nor is the effect of CR on other vascularized allografts yet known.

Preservation of Individual Organs For Clinical Transplantation

At present, vascularized grafts are most commonly and optimally preserved by static hypothermic storage, which for kidney, liver, and pancreas provides effective storage and likelihood of immediate function for up to about 24 hours of storage. Extended storage times beyond this are more liable to poor early function. Perfusion preservation for kidneys is relevant to such extended storage and also for resuscitation of ischemically damaged organs from asystolic cardiac donors with significant warm ischemic damage.

Kidney

The effectiveness of kidney preservation is measured by the quality of early kidney function after transplantation. Factors affecting that quality of preservation include donor management, nephrectomy technique, warm ischemia time, total renal ischemia time, and the anastomosis time.

Prospective, randomized clinical trials using preservation times averaging around 24 hours have shown no significant differences in early function or in long-term survival between machine preservation and simple cold storage. Extended cold-storage times of up to 2 and 3 days, although still capable of giving satisfactory results, are accompanied by an increased frequency of acute tubular necrosis (ATN) as storage times lengthen beyond 24–48 hours. Although in experienced units, delayed initial function is compatible with good long-term results, the advantages of immediate good renal function with a maintained urine output and rapid correction of azotemia after transplantation are striking.

Clinical management is markedly facilitated, with earlier hospital discharge and earlier recognition and treatment of rejection episodes. Kidneys with early ATN, even when this is followed by adequate function, do less well in terms of long-term survival than do kidneys functioning immediately, raising the likelihood that ischemically damaged kidneys may be more prone to immunological rejection.

Thus, even with the kidney, in which dialysis can maintain patient survival in the face of poor early function, good early function remains the paradigm of clinical organ preservation. Simple hypothermic storage can give early function in the majority of grafts. Storage times over 48 hours are rarely obligatory, except when organs are transplanted over very long distances.

UW-derived solutions (HES-free UW) have given successful static storage for 5 days in dogs. The gold standard for cadaver kidney preservation thus remains UW. Comparable results are achievable by a variety of other solutions when preservation times do not exceed 24 hours. Thus, for shorter anticipated preservation periods and for living-donor transplantation, UW, sucrose, Collins' solution, hypertonic citrate, and Bretschneider's solution can give equivalent results.

Static hypothermic storage

The length of time that a kidney can be stored depends on both cooling to diminish metabolic activity, and thus oxygen requirements, and on use of fluids designed to preserve the intracellular milieu in the absence of a Na^+/K^+ pump. Cold flushing washes out the blood and enhances cooling rate. As surface cooling is too slow and freezing would produce irreversible intracellular damage, perfusion of preservation fluid at 2°–4°C is undertaken through the renal arteries. Cooling by flushing through the aorta *in situ* can reduce renal core temperature to below 10°C in 3–5 min, compared with 20 min for an equivalent reduction by surface cooling. The preferred storage temperature in rat and dog kidney preservation has been shown to be 0°C rather than 5°C or 10°C. Very rapid cooling has been shown to be detrimental to rabbit kidneys, but such rapid rates of cooling are not obtained in larger human kidneys.

A miscellany of preserving solutions of varying solute composition has been developed over the last 30 years. Solutions of extracellular electrolyte composition, such as plasma, saline, or Hartmann's solution, are unsuitable for prolonged (>12 hours) preservation but are often used to flush-cool kidneys *in situ* during multiple-organ procurement. These solutions may also be used for cooling kidneys before immediate transfer in living-donor transplantation but are not recommended because alternative solutions have been demonstrated to be markedly superior. Both the original colloid-containing, high-potassium UW and the modified HES-free, high-sodium UW without some adjuvants have been shown to be equally effective in kidney preservation.

The recommended practice for cadaver kidneys is to submerge the organ in iced saline slush after removal from the body and to complete the flush with cold-preserving solution at 0°C until the effluent is clear of blood. The kidney is then surrounded with the perfusing solution, placed in a sterile container — usually double plastic bags, and stored in ice in an insulated container.

Perfusion with normal saline would result in transmembrane influx of sodium, loss of potassium, and rapid destruction of cellular integrity. A number of fluids have been designed that retain cellular viability for many hours or days. Cellular edema is reduced by use of impermeant solute such as mannitol, sucrose, or raffinose. Intracellular pH is preserved with buffers such as phosphate, citrate, or lactobionate, and high concentrations of potassium prevent influx of sodium and water. Commonly used solutions are either isotonic and mimic the intracellular electrolyte composition, or hypertonic such as Ross citrate solution. Any of these solutions preserve cold-stored kidneys for up to 24 or 48 hours.

The UW solution is more complex and provides no clear advantage over other preservation solutions for kidneys stored for up to 24 hours, but it may be more effective thereafter.

Oxygenated perfusion

An alternative to cold storage of the perfused kidneys packed in ice is continuous machine perfusion. In essence, a pump is used to recirculate either oxygenated synthetic stable albumin-based solution or UW at temperatures between 5°C and 10°C. Good renal preservation for up to 72 hours has been shown, but continuous perfusion has the disadvantage of being labor intensive and is unnecessary if the kidneys are transplanted within 24 hours of donation.

Continuous perfusion preservation is only clinically used for the kidney and only at a few centers. The balance of evidence indicates that this method provides better kidney preservation. Early reports based on retrospective comparisons of machine perfusion and cold storage showed no significant differences in allograft function after transplantation. However, these studies may have been biased in favor of cold storage, as the practice at the time was to use machine perfusion for much longer cold-ischemia times. In addition, little information was included about the recipient patients and the institutions where they were treated. In prospective randomized trials, there was no statistical difference in the rate of delayed graft function, and when delayed graft function did occur, machine-perfused kidneys showed earlier recovery and higher 1-year graft survival, a beneficial effect of the machine-preserved kidney, and less acute tubular necrosis was shown in the cold-stored organs when the preservation period exceeded 24 hours.

Later trials have concluded that kidneys preserved by machine perfusion show better early function than preservation by cold storage, and less dialysis requirement, which may correlate with initial graft dysfunction and long-term graft survival. Early and long-term renal function in kidneys preserved with hypothermic Belzer-MPS solution and machine perfusion has been shown to be improved, and it is even advantageous after kidneys have been initially cold stored. Machine perfusion can provide longer successful preservation periods and may produce significant savings from reducing delayed graft function and dialysis.

One of the limitations of preservation by perfusion is that it becomes less reliable after 5 days when considerable damage to the vascular endothelium can occur. This may be a result of hemodynamic effects on the endothelium, loss of vital metabolites, derangement of calcium compartmentation, and accumulation of toxic products in the cell and in the perfusate. The restoration of cellular homeostasis also becomes more difficult after prolonged periods of hypothermic preservation.

Preservation periods were extended for canine kidneys to 6 days by a 3-hour period of normothermic perfusion (either *ex vivo* or *in vitro*) at the halfway point. These observations have been confirmed and extended in dog and rat kidneys; however, the underlying mechanisms are still obscure. Provided damage during the period of normothermic perfusion can be avoided, this method provides the means to restore metabolite deficits, reverse cold-inhibited enzyme systems, and metabolize accumulated toxic products. A period of normothermic perfusion with blood is also effective in "*rescuing*" kidneys damaged by a period of warm ischemia.

The addition of metabolic inhibitors (e.g., quinacrine, a phospholipase inhibitor, and protease inhibitors, relevant to mitochondrial function and calcium compartmentation) may also be of value in prolonging kidney preservation up to 7 days. Thus far, 1 week has marked the limit of successful kidney preservation by machine perfusion. UW, containing HES, has formed the basis of such extension of preservation times; the UW-derived solutions can experimentally preserve kidneys (with minimal warm ischemia) reliably for 5 days by ice storage after a simple flush and for 7 days by continuous machine perfusion. Hypothermic perfusion has no significant advantages experimentally or clinically over static hypothermic storage for 24-hour preservation of kidneys procured with minimal warm ischemia. However, the recent resurgence of interest in use of kidneys from non-heart-beating donors has stimulated renewed efforts to provide reliable portable perfusion technology.

Liver

Static hypothermic storage

The introduction of UW revolutionized liver transplantation. Previous solutions had given successful preservation for only 8–16 hours in canine, porcine, or rodent models. UW allowed reliable and near-100% successful preservation in dog and in rat livers for 24 hours. Extended preservation to 30–48 hours was also possible experimentally, but with less adequate initial function.

Clinical practice requires immediate function for liver grafts. At present, 80% or more grafts function adequately, but about 10% show initially poor graft function, and about 5% have permanent nonfunction of the graft, necessitating life-saving retransplantation. Factors influencing and responsible for poor initial function include fatty change in the donor liver, older donor age, preliminary liver reduction surgery, and longer cold storage. Although storage for 24 or 30 hours can give good early function, it is thus preferable to keep storage times to under 12 hours for optimal results.

Benchmark clinical practice remains hypothermic storage using UW. Modifications without HES and with reversed Na^+/K^+ ratio are equally effective. When using high-K^+UW for preservation, a prevascularization rinse with a further washout solution is desirable. CRs are preferable to a Ringer's-lactate rinse. Important considerations, whatever the composition of the solution, are that a brief prevascularization rinse should be at room or body temperature and that a basic extracellular crystalloid rinse composition can be enhanced by addition of adenosine, allopurinol, desferrioxamine, and glycine. Other additions to UW have been pentoxifylline, chlorpromazine and calcium-channel blockers, prostaglandins, eicosanoids, and lazaroids. Future modifications are likely to incorporate these and other additives to storage and rinse solutions, together with glycine and reversed Na^+/K^+ electrolyte content.

Oxygenated perfusion

In the early 1960s, as liver transplantation was approaching clinical reality, Sicular and Moore investigated various methods of cooling and preserving organs by measuring glucose metabolism and carbon dioxide production. The researchers reported surprisingly good maintenance of function in livers perfused with acellular, oxygenated perfusate at 15°C. In 1963, Starzl used femoro-femoral, extracorporeal perfusion to preserve canine organs at 12°–15°C with autologous blood. Using this technique, function was limited to 6 hours for the kidney and 2 hours for the liver. In these experiments, perfusion was initiated after cessation of blood flow without flushing the organ *in vivo*; this created a period of warm ischemia. *In vivo*-isolated perfusion of the liver with oxygenated blood at 10°–18°C was used to preserve canine livers for up to 5 hours before successful transplantation. By 1967, the combination of continuous perfusion and hypothermic storage brought organ preservation to a new level. Using oxygenated plasma pumped in a pulsatile fashion at 8°–12°C, canine kidneys were successfully preserved for 72 hours, as described by Belzer and colleagues in a landmark paper. Since then, perfusion has recently undergone a revival, with livers being perfused in increasingly complex circuits for increasingly longer periods and at increased temperatures.

The use of machine perfusion applied to the liver presents a logistical concern because of the dual blood supply. Pulsatile perfusion of the artery alone using balanced salt solution at 4°C and 4 atmospheres pressure of oxygen gave marginal results in canines in 1967. Bretschneider perfused both the portal vein and hepatic artery with a perfusate composed of half autologous blood and half preservation solution under refrigerated, hyperbaric conditions. In this experiment, in canines, the organ was flushed *in vivo* before being placed on the machine and subsequently transplanted. All five animals survived with excellent hepatic function after 8–9 hours of machine preservation. Three of five survived after 24 hours of preservation. The method was successfully applied in seven patients, with periods of liver preservation lasting from 4 to 7 hours. All seven patients survived the first postoperative week. This effective technique never gained widespread acceptance on account of logistical constraints, but it marked the first successful machine preservation of the liver and emphasized the importance of perfusing the portal vein and flushing the organ *in vivo*.

Subsequent introduction of Collins' solution and successful preservation with static HIP (hypothermic ischemic preservation) moved the clinical focus away from OP (oxygenated cold perfusion). Clinical priorities then moved toward the logistic advantages of HIP, which enabled rapid and economic distant procurement and transport of organs.

Belzer later focused his OP work on the liver, using the porcine liver with the same cryoprecipitated plasma and hypothermic temperatures already used in kidney perfusion. This method involved continuous portal flow and pulsatile arterial flow. Four of five animals survived transplantation for 7 days after 8–10 hours of preservation, but only 2 of 12 survived beyond 12 hours when the preservation period was extended to 24 hours. Similar results have been seen in canines. The 24-hour barrier of OP in livers was achieved in 1973 by reports of 24- and 48-hour preservations of the canine liver. The reported technique involved arterial and portal perfusion of hypothermic, oxygenated plasma with the addition of corticosteroids. Twelve of 19 dogs survived 5 days in the 24-hour group, and 2 of 19 survived in the 48-hour group.

Rat livers were perfused for 20–24 hours at 10°C with a perfluorocarbon emulsion in the perfusate. Long-term survival (88–370 days) was achieved in five of nine rats in the fluorocarbon emulsion group, whereas the longest survivor in the control group lasted 5 days after transplantation. An advantage of machine perfusion over HIP is the ability to provide a continuous supply of the substrates necessary for cellular functions. These results emphasize the need to enhance the benefits of continuous perfusion with the addition of oxygen.

The best results to date in liver preservation by machine perfusion were obtained by Belzer's group, which used only the portal vein to perfuse canine livers for 72 hours at 5°C. This study showed the feasibility of using OP to achieve substantially longer preservation periods than can be achieved with HIP.

Heart

A shortage of organs, a relatively short acceptable ischemic time, and the serious perioperative consequences of inadequate preservation, as evidenced by a primary organ failure rate of 25% of early-recipient deaths, underline the importance of optimal hypothermic preservation techniques, in addition to rigorous selection criteria for donor hearts.

Experimental studies have demonstrated satisfactory preservation of cardiac allograft function after 24 hours of storage, which, unfortunately, has not translated into successful extension of preservation times in man. Current clinical preservation techniques generally permit a safe ischemic period for cardiac allografts of 4–6 hours. The disparity between experimental and clinical studies may be attributed to several factors. First, it is imperative that cardiac allografts function immediately on implantation, achieving near-normal functional capacity in the early postoperative period. Second, unlike the kidney and liver, the heart poorly tolerates depletion of ATP. Third, the status of the human donor allograft at explantation is an additional factor not accounted for in laboratory studies, with virtually all experimental studies performed using allografts from normal donor animals that do not take into account baseline injury before explantation. It is now clear that significant injury can result from suboptimal management of the donor before allograft procurement. Furthermore, with the increasing acceptance of marginal and older donors, in an effort to boost the number of available organs, there is an unavoidable increase in baseline coronary artery disease. Optimal donor selection and management must now be considered integral components of allograft preservation.

Preservation injury in the heart is multifactorial. Factors contributing to the severity of post-operative allograft dysfunction include insults associated with suboptimal donor management, hypothermia, ischemia-reperfusion injury, and depletion of energy stores. It has been suggested that organ preservation commences with meticulous treatment of the donor. Hemodynamic instability, loss of thermal regulation, and metabolic derangements accompanying brain death complicate the care of these patients Electrolyte imbalances exacerbate allograft edema and predispose the donor to dysrhythmias. Microinfarctions develop as a result of hypotensive and hypertensive episodes. It is important that aggressive efforts are taken to minimize these sequelae of brain death to ensure the potential suitability of the organ for transplantation and to decrease the likelihood of early and late dysfunction of implanted allografts. Selection and management of the donor candidate is at least as important as the preservation technique used.

The cardiac graft is further subject to injury at procurement. Excessive handling of the allograft during procurement may predispose the manipulated tissues to more severe preservation injury. Maintenance of allograft hypothermia and avoidance of extended exposure to air are also important for optimal preservation. Hypothermia is the cornerstone of current organ preservation. Unfortunately, it also contributes to the untoward cellular changes characteristic of *ex vivo* organ storage. The hypothermic inactivation of the Ca^+ ATPase and Na^+/K^+ ATPase results, in part, in the sequestration of calcium and the leakage of water across the sarcolemma and into the cellular organelles, respectively. Cardiac muscle is also very sensitive to depletion of energy stores during preservation. Static cold storage inevitably results in the exhaustion of high-energy phosphate stores. Without exogenous substrates for energy metabolism, the cardiac allograft must resort to finite glycogen and lipid stores, contributing to the limitations of safe storage times.

The predominant histological changes associated with heart preservation injury are myocardial edema bands and ischemic contracture. Contractures result from energy depletion and oxygen free-radical-mediated injury to intracellular organelles, with subsequent disturbance in calcium homeostasis. Mitochondrial damage exacerbates the progressive depletion of energy stores by further reduction in ATP availability. Altered membrane permeability and a reduction in Ca^+

ATPase activity in the sarcoplasmic reticulum, in conjunction with this decrease in high-energy phosphates, results in sluggish resequestration of calcium and, in turn, prolongation of diastolic relaxation. Cytosolic free calcium accumulates and binds with tropomyosin, creating a high-affinity state between actin and myosin. With ATP depleted, the actinomycin cross-bridges are unable to be broken, resulting in irreversible myocardial contractures. The resultant allograft dysfunction often presents as diastolic stunning and reduced ventricular compliance in the early postoperative period. Endothelial cell injury accompanying the global parenchymal insult may result in coronary vascular dysfunction. Depositions of activated leukocytes and platelets in capillary beds of the donor heart contribute to the development of the low-reflow phenomenon occasionally observed at reperfusion.

Medical management of cardiac donors is an integral part of organ preservation that is complicated by the complex physiologic phenomenon of brain death and the need to coordinate procurement with other organ transplantation teams. Optimal care requires that the donor be treated as any other patient in an intensive care unit, with invasive hemodynamic monitoring, ventilatory support, and meticulous attention to intravascular volume status and electrolytes, to avoid irreversible damage to the allograft, precluding transplantation.

Despite more than 2 decades of investigation, no single preservation regimen of the heart has yielded consistent, clinically superior preservation of allograft function. Hypothermia remains the most important component of organ preservation and the only factor universally considered to be essential, as it reduces cellular oxygen consumption, metabolism, and activity of hydrolytic enzymes; increases allograft tolerance to ischemic insults; and stabilizes lysosomal membranes. Because cellular metabolism continues during cold storage — albeit at a greatly reduced rate — hypothermia can provide only a finite amount of protection before energy stores are eventually depleted, and toxic metabolites accumulate as anaerobic metabolic pathways commence.

Static hypothermic storage

The addition of a single flush of cardioplegic solution before static hypothermic storage significantly improves myocardial protection over hypothermia alone. At the time of organ procurement, a bolus of cardioplegia solution is administered in standard fashion with a 14-gauge catheter proximal to the aortic cross-clamp. Rapid topical cooling of the heart is achieved with cold saline poured into the pericardial well. After explantation is complete, the allograft is sequentially placed in two sterile bags, each filled with cold saline; a saline-filled air-tight container; and finally a standard cooler of ice for transport. This simple technique provides excellent short-term myocardial protection and is the primary method used by more than 90% of transplant centers. This simple technique of static hypothermic preservation provides excellent protection for currently accepted safe ischemic times. Target temperature for heart allograft storage is institution dependent and generally ranges between 0°C and 7°C.

Experimental evidence indicates that the optimal temperature for myocardial protection approximates to 4°C — the temperature of melting ice. This evidence supports the use of the simple, ice-slush static storage system. Because the temperature of the storage solution likely falls below 4°C, and few centers actually monitor the temperature, the allograft should be immersed in sufficient saline to avoid cryogenic injury. Supercooling or freezing techniques using subzero temperature storage are being explored to extend the period of safe ischemic times. Efforts are focused on prevention of intracellular crystal formation and optimal thawing protocols. Cold saline is the most frequently used storage medium. In the Papworth heart donor survey, storage in any nonsaline solution was correlated with a greater than twofold increase in mortality.

Controlled studies are necessary to confirm this finding. The time necessary to perform implantation of the heart may account for a significant amount of the total ischemic time. During this period, the allograft may be warmed by direct thermal transfer from adjacent thoracic structures as well as from pulmonary venous drainage. Continuous aspiration of pulmonary venous return may be achieved by insertion of a vent into the left atrium through the right superior pulmonary vein or left atrial appendage. After completion of the posterior left atrial suture line, topical cold saline irrigation is initiated, and the patient is oriented in a left-side-down, head-up position to allow drainage of the saline away from the operative field and maximal exposure of the left and right ventricles.

It is essential that the dependent portions of the allograft remain immersed in cold saline and that any exposed regions be covered with cold saline-soaked gauze pads. To shorten the allograft ischemic period, some centers perform the aortic anastomosis before that of the pulmonary artery, permitting completion of implantation with the cross-clamp removed. However, others suggest that the additional 10 minutes of ischemia necessary to perform the pulmonary anastomosis is a relatively short amount of time to exchange for greatly facilitating this important anastomosis.

Oxygenated perfusion

In the search for methods to extend preservation times, the efforts of some early investigators were directed toward developing a successful continuous-perfusion system for the explanted heart. The theoretical advantages of perfusion preservation over static storage are the presence of a constant supply of oxygen and substrate for basal metabolic demands and the washout of toxic byproducts of metabolism. Unfortunately, these potential benefits were often overshadowed by the progressive increase in myocardial edema, in addition to the technical problems inherent in a complex perfusion apparatus. It is unlikely that human cardiac allografts preserved with current perfusion technology would be able to sustain recipients if transplanted orthotopically. Microperfusion techniques were proposed to reduce interstitial edema and associated diastolic dysfunction accompanying continuous perfusion. They are characterized by perfusion at low pressures to attenuate the intravascular hydrostatic component of Fick's equation and thus reduce fluid extravasation. Perfusate oncotic pressure may be enhanced with the addition of solution impermeants to minimize edema further.

A compromise between single-flush and continuous perfusion cardioplegic delivery is a multiple-flush technique extended over the course of the ischemic interval. Although experimental evidence suggests some improved parameters with intermittent perfusion, clinical application has remained limited.

Composition of perfusate and storage solutions

Historically, many solutions have been used for myocardial preservation, with varying success. Because the optimal perfusate remains elusive, solutions of widely different compositions are currently championed by their advocates.

Cardioplegic solutions

To date, cardioplegic solution has not been an independent predictor of mortality in multivariant analysis. Current intracellular solutions for cardioplegia include EC, Belzer's UW, and Bretschneider (HTK) solutions. Early work with extracellular solutions as perfusate-storage media failed because of the loss of endothelial integrity and the development of significant cellular edema; however, addition of impermeants eventually permitted their successful use clinically. Stanford, Hopkins, and St. Thomas Hospital solutions are the representative extracellular cardioplegic solutions. Of electrolytes included in cardioplegic-storage solutions, the optimal concentration of potassium and calcium has stimulated the most controversy. In selected situations, high concentrations of potassium have been associated with cellular injury, particularly of the coronary endothelium.

Nevertheless, most investigators agree that convincing evidence demonstrating significant injury of the hypothermic cardiac allograft secondary to currently available hyperkalemic solutions is lacking. Proponents of the addition of calcium warn of the possibility of the calcium paradox, with acalcemic solutions (particularly in the setting of a low- sodium preservation medium); in contrast, calcium overload is intimately involved in the development of ischemic contractures in the myocyte. Experimental evidence indicates that only micromolar amounts are necessary to prevent the rare calcium paradox phenomenon, thus reducing the risk of contributing to contracture formation.

Solution additives

Recognizing that oxygen-derived free radicals are likely the critical mediators of ischemic-reperfusion injury, pharmacologic interventions to neutralize these cytotoxic molecules were a natural extension of heart preservation research. Administration of exogenous free-radical scavengers is believed to assist the recipient's intrinsic antioxidant defenses in the metabolism of the oxygen intermediates to benign molecules. Although the addition of scavengers or chelators to the preservation regimen of heart allografts has resulted in a reduction in allograft edema and improved post-transplant function in numerous studies, their use remains controversial because of unsatisfactory or inconsistent results when administered clinically. Impermeants are high-molecular-weight molecules that do not readily penetrate cell membranes.

Mannitol, lactobionate, and raffinose are examples of impermeants that act as extracellular osmotic agents. They purportedly reduce hypothermia-induced cellular edema in allografts by counteracting the intracellular osmotic pressure imparted by protein and other cytosol-confined molecules. Other impermeants such as hydroxyethyl starch and dextran act as oncotic agents and help prevent the development of allograft interstitial edema. The need to include impermeants in static storage solutions remains debatable. Because storage of organs at 4°C does not completely arrest metabolism, it is not surprising that metabolic substrates are common additives in perfusate-storage solutions. Investigations of metabolic substrate supplementation with adenosine, pyruvate, glutamate, and aspartate have dominated the myocardial preservation literature.

ATP may be repleted from adenosine through the purine salvage pathway and from these amino acids through the citric acid cycle. Although the biochemical rationale for provision of substrates in perfusate storage solutions appears theoretically sound, it is still unclear whether exogenous substrates can be taken up and used by the hypothermic cells of the allograft and, moreover, whether their addition significantly improves outcome.

Steroids (methylprednisolone, dexamethasone) have been added to preservation solutions for their general antiinflammatory effects. Because ischemic injury to the allograft is partially mediated by a calcium influx, the addition of calcium channel blockers has also been proposed. Finally, myocardial ischemic injury may be ameliorated by the administration of potent inhibitors of lipid peroxidation, called *lazaroids*.

Many issues regarding perfusate storage solutions remain unresolved, and the spectrum of diverse experimental models complicates reliable comparison of results in the literature. Perhaps, as some investigators have theorized, if allograft ischemic periods are extended in the future, electrolyte composition and pharmacologic additives will become increasingly more important clinically. However, despite the vehement claims of investigators worldwide, there is no compelling clinical evidence at this time that any solution is clearly superior to the others for cardiac preservation when used within the current 6-hour ischemic limit.

Non-heart-beating donor hearts

Because of the critical shortage of allografts, the use of non-heart-beating donors is now being explored. The period of warm ischemia before procurement further complicates preservation strategies. Furthermore, preparation of the procurement team is difficult, with the unpredictability of the time of death. Although some organs may tolerate the warm ischemic insult, it remains uncertain whether reanimation of hearts from these donors will yield allografts adequate for successful clinical transplantation.

The critical components of successful cardiac allograft preservation include meticulous donor management, hypothermia, and reperfusion modification. Continued research will be necessary to explore systematically the complex mechanisms mediating donor allograft injury and the many variables affecting the elusive optimal preservation strategy of cardiac allografts.

Lung

The quality of lung preservation affects perioperative allograft function, as poor allograft preservation results in increased morbidity and mortality following lung transplantation and could upregulate the expression of histocompatibility antigens, potentially resulting in increased allograft rejection.

In addition, better lung preservation techniques could extend allograft ischemia times, resulting in increased numbers of organs available for transplantation, better procurement strategies, and prospective human leukocyte antigen matching. These factors have led to the continued search for better preservation modalities to decrease the extent of ischemia-reperfusion lung injury.

Since the first human lung transplantation procedure was performed in 1963, much progress has been made in the field of lung allograft preservation. Initially, lung transplantations were performed locally, with the donor transported to the recipient center. This resulted in minimal graft ischemic times and a decrease in the possibility of primary nonfunction, which is not well tolerated in the lung. Despite extensive experimental research, graft preservation beyond 2–3 hours was difficult to obtain before 1980.

A variety of techniques were subsequently developed to allow for the preservation of lung grafts over greater distances and to allow for graft ischemia times of 4–6 hours. These techniques have included simple hypothermic atelectatic immersion, *ex vivo* normothermic heart–lung autoperfusion, donor core cooling, and pulmonary artery flush. The first two of these methods are of historical interest only and are no longer used in clinical practice.

Hypothermic atelectatic immersion

Hypothermic atelectatic immersion was the initial method of preservation used by the Toronto Lung Transplant Group. Satisfactory clinical results after 5.5 hours of ischemia were achieved. This method was used for both single- and double-lung transplantation with good results. However, the negative effects of atelectasis later led to the abandonment of this technique in favor of pulmonary artery flush, which allowed longer ischemic times with more predictable postoperative graft function.

Oxygenated perfusion

Ex vivo normothermic heart–lung autoperfusion was also used successfully in the early clinical experience of lung transplantation. Originally described by the physiologists Martin and then Starling in the early 20th century, it was adapted in the 1960s for the preservation of heart grafts. The Pittsburgh group then subsequently adapted this technique for the preservation of heart–lung grafts, with good results. This method was later abandoned because of early deterioration of lung grafts from presumed neutrophil sequestration, resulting in uncontrolled pulmonary hypertension, hypostatic pulmonary edema, and graft failure. The procedure was also technically complex and required mechanical ventilation of lung grafts via the donor trachea during transport, preventing the sharing of intrathoracic organs among different transplantation centers. This technique is no longer used clinically, but it does provide a good model for the study of lung function in the laboratory.

Static hypothermic preservation

Static preservation with hypothermia is the mainstay of current lung preservation of which two variants dominate the clinical field. To ensure complete cooling of the lung, donor core cooling has been employed before organ removal with good results. The other common technique is similar to that employed in the liver and kidney and consists of organ perfusion with cooled perfusate *in situ* before removal through the pulmonary artery.

Donor core cooling

Yacoub and colleagues were the main proponents of donor core cooling on cardiopulmonary bypass, especially for combined heart–lung transplantation. The donor core cooling technique has not undergone major changes since its introduction into clinical practice in 1985. It is estimated that more than 50% of the world's heart–lung transplantation procedures have been performed with organs preserved in this fashion. The technique results in controllable hypothermia with protection of other thoracic and abdominal organs. The perfusate is blood, which is an effective pH buffer, allows for tissue metabolic activity, and contains free-radical scavengers.

The pulmonary parenchyma is also protected against embolism during perfusion because its only inflow comes from the bronchial arterial system. Despite these advantages, however, there are several disadvantages. First, the technique is cumbersome, requiring a portable cardiopulmonary bypass unit (roller pump, perfusion circuit, heat exchanger, and oxygenator) in addition to the appropriate skilled personnel. Second, cooling of the lungs is slow and inefficient, requiring large amounts of ice and a prolonged period of time to achieve the necessary temperature drop. Third, the use of cardiopulmonary bypass results in deleterious effects on pulmonary function both from a direct ischemic insult and from the activation of various inflammatory pathways, which results in increased alveolar capillary permeability. Fourth, animal data have indicated the presence of reduced oxygenation capacity in lungs preserved with this method compared with the pulmonary artery flush technique, although postoperative pulmonary vascular resistance was lower with core cooling, suggesting better preservation of the pulmonary vasculature.

A study in bovine lung allografts, however, revealed that the addition of recipient and donor leukocyte depletion to donor core cooling resulted in improved post-ischemic lung function beyond that provided by donor core cooling alone, resulting in good lung function after 24 hours of ischemia. Given these concerns about donor core cooling for lung allograft preservation, the pulmonary artery flush technique has become the accepted preservation method used in lung transplantation.

Pulmonary artery flush

As in kidney, liver, and heart preservation, the most common technique for lung preservation is simple flushing with a cold solution. In the lung, this is followed by hypothermic immersion in a semi-inflated state. Although at one point continuous flushing of lung grafts was investigated, this research has since been abandoned because of increased pulmonary hypertension, decreased lung compliance, and worsening pulmonary edema.

On the basis of initial experimental and clinical work at Stanford University, the pulmonary artery flush technique has been adopted by the vast majority of lung transplantation centers worldwide. Initial work at Stanford University centered on the use of cold potassium cardioplegia to flush both the pulmonary and the coronary arteries for combined heart–lung transplantation. Subsequently, cold EC solution was modified with magnesium sulfate and 50% dextrose. This modified solution was used to flush the pulmonary arteries, and cardioplegia was used to simultaneously perfuse the coronary circulation. Subsequently, modified EC became the solution of choice for lung preservation.

The technique of pulmonary artery flush is straightforward and simple and allows for the rapid cooling of both lungs. The extent of cooling, however, depends on the volume used and on the distribution of this perfusate within the lungs. Increasing both the flow rate and the total volume of perfusate used resulted in significant improvement in postoperative graft function.

This method has been adopted into current clinical practice, in which a total perfusate volume of 60 mL/kg of body weight is typically infused over a 4-minute period. The technique also removes potentially harmful blood products such as platelets, leukocytes, and complement. Embolization into the pulmonary microvasculature is possible, however, and care must be taken to avoid this complication. The perfusion pressure should be kept low, and the flush temperature should be kept at 4°C as much as possible to decrease other potential complications.

Although the majority of centers flush the lungs in an antegrade fashion, an interesting variation is a retrograde flush of the lungs via the left atrium. In a porcine model of lung preservation with modified EC, radiolabeled albumin was used to assess distribution within the tracheobronchial tree. Retrograde flush resulted in higher radiolabeled albumin counts compared with antegrade flush. Results were also encouraging in a study of 21 consecutive lung transplant recipients who received lungs preserved via the retrograde route. Whether this technique is significantly different from antegrade pulmonary artery flush when applied on a more widespread basis remains to be seen, although the preliminary evidence is encouraging.

Composition of preservation and flush solutions

During the period of graft ischemia, the activity of the Na^+/K^+ ATPase pump is decreased, leading to cellular swelling. To counteract this problem, two types of preservation solutions have been developed: intracellular solutions, such as EC and UW, and extracellular solutions, such as Wallwork and low-potassium dextran solutions.

Among the intracellular solutions, the most commonly used worldwide is the EC modified by the addition of magnesium sulfate and 50% dextrose. The addition of glucose is based on animal studies showing better immediate lung function and higher levels of glucose metabolites in lungs preserved with a glucose-containing, low-potassium, dextran solution, compared with lungs preserved with only a low-potassium dextran solution. The osmolarity of modified EC solution is slightly higher than that of serum. This has resulted in the preservation of lung grafts for up to 8–10 hours, although the upper limit is difficult to ascertain, given the many confounding variables affecting graft function in the clinical setting.

Another intracellular solution in clinical use in some centers is the UW. Although its use in the preservation of liver, kidney, and pancreas grafts is well documented and results in the extension of preservation times for these organs, its use in lung transplantation is still limited. It contains several components, the actual roles of which in ameliorating ischemia-reperfusion injury are not well understood. The K^+ concentration in UW (120 mmoL) is similar to that of EC (115 mmol/L). This high concentration of potassium ions can result in pulmonary vasoconstriction and nonuniform flushing of the grafts if no pulmonary vasodilator, such as a prostaglandin bolus infusion, is used before flushing.

A complicating factor for UW is that its efficacy may be related to its shelf life, in contrast to EC. This is probably related to one of its components, reduced glutathione, which is a neutralizer of oxygen free radicals and lipid peroxides. Following prolonged storage, glutathione is converted to oxidized glutathione. However, the effect of this change on lung preservation and post-operative graft function is not clear.

In general, it is thought that UW is as effective for lung preservation as the modified EC. There is some evidence to indicate that longer periods of preservation may be possible with UW. The introduction of a low-potassium UW has resulted experimentally in improved gas exchange in isolated rabbit lungs. Whether this result is true in the clinical setting has yet to be determined. The majority of lung transplantation centers worldwide, however, continue to use modified EC solution for graft preservation primarily because of its longer duration of use in lung preservation and its lower cost and complexity in comparison with UW.

An alternative to intracellular solutions is an extracellular solution developed by the Papworth group, known as Wallwork solution. It is based on third-party cold donor blood that is modified by the addition of various substrates such as epoprostenol (formerly prostacyclin), buffers, and human proteins, achieving a flush hematocrit of approximately 10%. Reports of its clinical use are limited, although preservation times of up to 4 hours have been achieved. In a comparison study of Wallwork solution and modified EC, conducted in a canine lung transplantation model, lungs preserved with Wallwork solution demonstrated lower compliance and worse postreperfusion pulmonary edema. Contradictory results were obtained when an *ex vivo* rat lung allograft model was used. In this latter study, ischemia reperfusion lung injury was most decreased using Wallwork solution, compared with EC, UW, and low-potassium dextran solution. In another experimental transplantation model, both low-potassium and high-potassium lactobionate solutions gave equivalent lung function after 24 hours of cold preservation. Further improvements in lung preservation were obtained when epoprostenol or the nitric oxide precursor L-arginine were added to Wallwork solution. Whether these results are similar in the setting of human lung transplantation has not been determined.

Another extracellular solution is low-potassium dextran solution, initially derived from animal experiments and then applied to clinical practice. The use of dextran 40 in this solution confers potential advantages, such as its role as a free-radical scavenger, its anticoagulant properties, and its beneficial effects on tissue oncotic pressures. Whether these benefits render low-potassium dextran solution superior to EC solution is controversial. In canine left lung allograft transplantation, low-potassium dextran solution was compared directly with modified EC solution after donor pretreatment with alprostadil (formerly prostaglandin E1) infusion. Immediately following transplantation and at 3 days postoperatively, there were no significant differences in gas exchange or hemodynamic function between the two experimental groups. Another study, however, showed beneficial effects of extracellular solutions compared with intracellular solutions, using a perfused and ventilated rabbit lung model. Both low-potassium dextran and blood-low-potassium dextran solutions provided superior pulmonary function compared with that of modified EC preceded by a bolus of alprostadil. Given this controversy, EC continues to be used for lung allograft flushing, typically after a bolus of prostaglandins is inserted into the pulmonary circulation.

A variety of pharmacologic additives have been investigated in animal models of lung preservation and transplantation, although the translation of these laboratory results into clinical practice remains slow. The additives that have been shown to ameliorate post-transplantation ischemia-reperfusion injury in various experimental models include the following: prostaglandins such as alprostadil and epoprostenol; oxygen free-radical scavengers such as catalase, glutathione, superoxide dismutase, allopurinol, dimethylthiourea, lazaroid, and deferoxamine; calcium channel blockers such as verapamil, diltiazem, nifedipine, and nicardipine; lidocaine; complement inhibitors such as nafamostat mesylate and the soluble form of complement receptor 1; platelet-activating factor inhibitors; pentoxifylline; surfactant;

monoclonal antibodies directed against intercellular adhesion molecule 1, CD11b antigen, or CD18 antigen; leumedins such as NPC 15669 and NPC 18915; sialyl Lewis^x analogs such as CY 1503; and NO donors such as inhaled NO, nitroglycerin, nieorandil, glyceryl trinitrate, and nitroprusside. Leukocyte depletion by filtration has also been shown to decrease reperfusion injury in experimental models. More recently, gene therapy of donor lungs to ameliorate post-transplantation ischemia-reperfusion injury has been investigated in animal models. This strategy may hold promise as a future therapeutic option in lung preservation.

A few transplantation centers have begun adding nitrates to the flush solution in clinical lung transplantation to decrease the extent of subsequent acute lung dysfunction. In an *ex vivo* reperfusion model of rat lung grafts, the addition of glycerol trinitrate to the flush solution improved oxygenation and lung graft integrity, resulting in better allograft preservation than that seen with the addition of epoprostenol. Another NO donor is sodium nitroprusside, which has been studied in a canine left lung allograft transplantation model. The administration of nitroprusside to the flush solution and during allograft reperfusion resulted in reductions in pulmonary vascular resistance and myeloperoxidase activity, a measure of leukocyte sequestration. In addition, the nitroprusside group had improved lung allograft function and blood flow, correlating with reductions in the extent of ischemia-reperfusion injury. Similar results were obtained when sodium nitroprusside was directly infused into the pulmonary artery in an isolated, ventilated, whole blood–perfused rabbit lung model. This resulted in improvements in pulmonary hemodynamics, oxygenation, compliance, and lung edema formation, without significant systemic hypotension. In a rat lung transplantation model with contralateral right pulmonary artery ligation, the addition of nitroglycerin to lactated Ringer's irrigation resulted in improved oxygenation, pulmonary blood flow, pulmonary vascular resistance, and survival compared with controls. Nitroglycerin has been incorporated into a new pulmonary artery flush solution, resulting in better preservation of canine lung allografts.

Given these beneficial effects of NO donors in lung allograft transplantation, sodium nitroprusside has started to be added to the flush solution. Initially, it was added to the flush solution at the donor harvest site just before pulmonary artery flushing. Since then, sodium nitroprusside has been incorporated into the modified EC flush solution. Another pharmacologic intervention that has been studied in the clinical setting is the inhibition of a complement, using a soluble complement receptor 1 inhibitor called TP10. A multicenter trial enrolled 59 patients who were randomly assigned to TP10 administered before reperfusion or to placebo. TP10 resulted in significant increases in early extubation following lung transplantation, although there were no statistical differences with respect to total time spent on the ventilator or total intensive care unit time. Increased benefit was observed in patients who had experienced both ischemia-reperfusion injury and cardiopulmonary bypass. Although these results are encouraging, more work needs to be done before complement inhibition becomes a clinical reality.

Additional factors affecting lung storage

Given the inherent differences of the lung compared with other solid organs available for transplantation, various physiologic factors during the storage phase have been studied to ameliorate subsequent ischemia-reperfusion injury. These factors have included the storage temperature of the graft, the state of graft inflation, and the oxygen concentration of the graft.

Graft temperature

Hypothermia is the cornerstone of lung preservation, although the exact degree of hypothermia has been a matter of controversy. The beneficial effects of hypothermia arise mainly from decreased cellular metabolic rates and slower substrate depletion rates. However, profound hypothermia can result in abnormal cellular membrane transport functions and calcium homeostasis. In addition, the activity of the Na^+/K^+ trans-membrane pump is inhibited, resulting in decreased plasma membrane integrity. There is experimental evidence in an isolated rabbit lung model and a canine left lung allotransplantation model indicating that the optimal temperature for lung preservation may be 10°C rather than 4°C. This has not been translated into clinical practice, however, because of the cumbersome nature of maintaining grafts at 10°C, which would require a portable temperature regulator and cooling cabinet. Following procurement, the vast majority of transplantation centers store lung allografts in ice-cold saline or slush during transport. One potential complicating factor with the use of this technique is the inhomogeneous cooling that results following cold immersion because of the buoyancy of the inflated lung. An alternative method, investigated in the laboratory, is the use of ambient compressed air with dry ice for the preservation of lung allografts. Favorable results were obtained in nonflushed canine lungs preserved in this way compared with immersion in iced slush. The translation of these results into clinical practice awaits further investigation before widespread acceptance.

Graft inflation

The initial clinical experience in lung transplantation employed grafts preserved by atelectatic cold immersion, for which the graft is completely deflated with the use of a bronchial blocker. With the introduction of the technique of pulmonary artery flush, it became apparent that the storage of grafts in an inflated state provided better graft preservation and decreased postoperative pulmonary edema. Experimental evidence in a canine lung transplantation model demonstrated that hyperinflated lungs resulted in more reliable post-transplantation graft function compared with grafts inflated at low lung volumes. Hyperinflation applied clinically, however, resulted in increased acute allograft dysfunction. Lungs stored at normal lung volumes, in contrast, demonstrated satisfactory allograft function and minimal pulmonary edema during the postoperative assessment period. Current practice is 100% oxygen end tidal volume inflation until implantation.

Storage oxygen concentration of the graft

The lung is unique among solid organs available for transplantation in that it is capable of maintaining aerobic metabolism after the cessation of cardiac activity. During extended periods of preservation, infra-alveolar oxygen and intracellular glucose allow the hypothermic lung to continue aerobic metabolism and preserve ATP levels. The lung may also be capable of maintaining aerobic metabolism for short periods in non-heart-beating cadavers, as shown by adequate arterial oxygenation in canine lungs harvested from donors several hours after death. For these reasons, it is common in clinical practice to inflate harvested lung grafts with oxygen, although there is still controversy about the exact oxygen concentration that is desirable. Experimental evidence has been contradictory. Ventilation with 100% nitrogen has been shown to be superior to room air or 100% oxygen in isolated canine lungs. In isolated rabbit lungs, however, ventilation with 100% oxygen was superior to room air, which was superior to 100% nitrogen. In an *in vitro* gravimetric rabbit lung model, graft inflation with 100% oxygen or 100% nitrogen increased the pulmonary capillary filtration coefficient, in comparison with storage at room air. In further studies in the same model, this increase in the capillary filtration coefficient correlated with increases in end products of lipid peroxidation. This was counteracted by dimethylthiourea, a free-radical scavenger, indicating that oxygen free-radical injury occurs during ischemia and not only during reperfusion. The use of hyperbaric oxygen has also been investigated, yielding favorable results, although the procedure is technically cumbersome and the experiments were often not controlled. Current practice is 100% oxygen until implantation.

Pretreatment

Treatment of the donor before pulmonary artery flush is used in virtually all lung transplantation programs to improve pulmonary artery flush distribution and decrease subsequent ischemia-reperfusion injury in the allograft. The high potassium concentration in intracellular flush solutions, pulmonary artery distention during graft flushing, and cold temperatures of the perfusates all result in reflex pulmonary vasoconstriction. This can theoretically result in nonuniform distribution of the flush solution and worse organ preservation. Prostaglandin administration into the donor lung before flush solution administration has been used to decrease this injury.

Bolus administration of prostaglandins into the pulmonary circulation before pulmonary artery flush has been routinely used in clinical practice to counteract reflex pulmonary vasoconstriction arising from the use of intracellular preservation solutions. Alprostadil is used in North America, and epoprostenol is used in Europe. These two prostaglandin formulations are considered clinically equivalent with respect to their effects on allograft preservation in lung transplantation.

The main benefit from the use of prostaglandins is pulmonary vasodilation, which counteracts reflex pulmonary vasoconstriction and allows for more uniform lung flushing and better organ preservation. Prostaglandins are also potentially beneficial for a variety of other reasons: decreased leukocyte sequestration, inhibition of platelet aggregation, prevention of lysosomal enzyme release and superoxide anion production by neutrophils, better cytoprotection, immunosuppression, and decreased vascular permeability. Whether these other effects of prostaglandins further contribute to improved lung preservation remains controversial. A study of the mechanism of action of alprostadil in rat lung transplantation indicated that vasodilation by itself was insufficient to enhance lung protection. Alprostadil enhanced lung preservation in that model by stimulating cyclic adenosine monophosphate-dependent protein kinase, thereby promoting non-vasodilatory mechanisms of lung allograft preservation. A variety of experimental lung transplantation studies have shown beneficial effects to donor pretreatment with prostaglandins before pulmonary artery flushing. The beneficial effects of alprostadil pretreatment were also seen when canine lung allografts were flushed retrograde with alprostadil followed by UW or infused during the first 6 hours of reperfusion, supporting the use of alprostadil in the clinical setting following lung transplantation, with similar results reported using epoprostenol or Iloprost, a stable synthetic derivative of epoprostenol with a longer half-life and fewer systemic hypotensive effects. The vast majority of lung transplantation centers use prostaglandins, mainly for their effects on pulmonary vasodilation, resulting in a high-volume, low-pressure pulmonary artery flush. Clinically, alprostadil is administered via the pulmonary artery, as a single bolus injection into the main pulmonary artery before the administration of the flush solution.

Treatment of the recipient after implantation has also been shown to be beneficial. In addition to standard immunosuppressive and antibiotic therapy in the postoperative period, various therapies have been used clinically to varying degrees to ameliorate the extent of graft ischemia-reperfusion injury. These have included the use of prostaglandins, high-dose methylprednisolone, and inhaled NO. Prostaglandins are also used clinically in the early postoperative setting to decrease reperfusion injury. In Europe, epoprostenol is administered until the systemic pressure is reduced by 30%, whereas in the United States, alprostadil is administered to control pulmonary artery pressures and reduce pulmonary vascular resistance.

Single-dose methyl-prednisolone has been added to most clinical protocols mainly because of its anti-inflammatory actions on activated neutrophils. Specifically, corticosteroids reduce lysosomal enzyme release and superoxide anion production by neutrophils, block complement-induced neutrophil aggregation, and inhibit arachidonic acid metabolism in granulocytes. Experimental evidence indicates benefits if corticosteroids are used in the donor before ischemia. Clinically, however, corticosteroids are usually administered as an intravenous pulse of methylprednisolone to the recipient immediately before reperfusion. An additional pharmacologic intervention used with marginal lung grafts at risk for reperfusion injury is inhaled NO in the post-transplantation period. Inhaled NO has a variety of potential beneficial effects in the lung. It is a potent yet selective pulmonary vasodilator, inducing relaxation of pulmonary vascular smooth muscle cells by activation of guanylate cyclase. NO has a variety of other actions inhibiting early neutrophil adhesion; preventing increases in microvascular permeability; inhibiting platelet adhesion, activation, and aggregation; and regulating the proliferation of vascular smooth muscle cells *in vitro* and *in vivo*. Clinically, inhaled NO has been shown to ameliorate acute lung dysfunction, although frequent monitoring of plasma methemoglobin levels is necessary to avoid toxicity.

Current practice in lung preservation is full heparinization of the donor, a bolus pulmonary artery infusion of alprostadil followed by inflow occlusion by ligation of the superior vena cava and cross-clamping of the inferior vena cava. Three liters of modified EC solution (added 50% glucose, magnesium sulfate and sodium nitroprusside) is administered, ice cold, into the main pulmonary artery via a large-bore catheter, usually a 14-gauge cannula. The perfusate is positioned 2 m above the ground to allow drainage by gravity. Perfusion pressures are usually 15–20 mm Hg. Simultaneous with the administration of flush through the main pulmonary artery, cardioplegia is infused into the ascending aorta after aortic cross-clamping. The right heart is vented by transecting the inferior vena cava proximal to its cross-clamp. The team harvesting abdominal organs is asked to cannulate the abdominal segment of the inferior vena cava to allow drainage of the perfusate off the table. The left atrial appendage rip is also cut to allow for drainage of the flush solution from the lungs.

Ventilation with 100% FIO_2 is administered manually throughout this phase to result in an even distribution of the perfusate within the lung parenchyma. The chest is flooded with iced saline, and cold effluent is allowed to collect in the pleural spaces to further topically cool the lungs. After excision of the heart and decannulation of the pulmonary artery and the ascending aorta, the lungs are extracted en bloc. They are moderately inflated with 100% FIO_2 to a final volume corresponding to normal end-tidal inspiration. An additional lung flush is occasionally administered at this time by retrograde infusion of modified EC solution into the pulmonary veins. Following separation of the intrathoracic organs, the grafts are transported under hypothermic conditions by placement in ice, which results in temperatures between 0°C and 1°C.

Ischemic times are typically less than 8 hours, although 10 hours is considered the upper limit. During implantation, the graft is protected from warm ischemia by topical cooling with iced slush. Immediately before reperfusion, a pulse dose of methylprednisolone is routinely administered to the recipient. Following reperfusion, intravenous infusion of alprostadil in the immediate perioperative period is administered for 24 hours. Methylprednisolone is continued intravenously twice daily for the first 3 days. Inhaled NO is administered at the first evidence of reperfusion injury.

Heart-Lung Block

Simple cooling has been used for short-term preservation for heart–lung transplants. Lungs were originally flushed with cold EC via the pulmonary artery immediately after induction of cardioplegia. Flushing should maintain pressure below that normally found in the pulmonary artery. Hyperinflation of the lung before storage was shown to be beneficial. Treatment of the donor with prostaglandin E1 followed by a hypothermic flush gave adequate 6-hour preservation.

Single lung transplantation is now increasingly common, and lung preservation using a EC flush is only successful for 5–6 hours. Progressive deterioration with increased pulmonary vascular resistance and decreased compliance occurs by 12 hours, with interstitial hemorrhage and edema by 24 hours. The disaccharide trehalose has been substituted for glucose in EC with advantage. Increasing experience with lung and heart–lung transplantation has demonstrated that the lung may be more robust in relation to its tolerance of extended storage than previously thought. UW and other solutions have given successful preservation for 24 hours. Additions to the preserving solution that have been found useful have included prostaglandin E1, pentoxifylline, verapamil, and glutathione. Equivalent results to UW have been found with a dextran-based, low-potassium solution.

Pancreas

UW was originally introduced to improve pancreas preservation. In the pancreas, it is paramount that preservation conserves the beta cells. Exocrine cell function can provide a marker of viability in the form of amylase excretion. Another important aim of preservation is to minimize the occurrence of ischaemia-induced acute pancreatitis, which increases morbidity after transplantation. Preservation of the pancreas proved difficult using Collins', citrate, and other solutions, and it was not until the advent of UW that reliable preservation for 24 hours was achieved experimentally and clinically. Experimental preservation has since been extended to 96 hours using UW with an added prostanoid inhibitor. Other variations of UW can give satisfactory cold storage, including removal of HES or its replacement with dextran, and combinations of sodium lactobionate with histidine.

The pancreas has also been studied in relation to the benefit of additional oxygen using a two-layer storage technique with a perfluorochemical adjacent to UW. In small animals, the technique of cold storage, whether surrounded by the preserving liquid or wrapped in moistened gauze, has influenced efficacy of preservation, perhaps related to the filamentous and nonencapsulated nature of the pancreas, resulting in a greater tendency for harmful edema and weight gain to occur during storage immersed in liquid. Whole-organ pancreas preservation needs to include the associated duodenal segment containing the duodenal papilla and pancreatic ducts.

Small Bowel

Small-bowel grafts can be transplanted alone or combined as organ-cluster grafts with liver or pancreas. Commonly, organ donors are treated with both OKT3 and antithymocyte globulin in an attempt to immunomodulate the potential allograft. An osmotic agent is not used to flush the intestinal contents, nor is any specific intestinal decontamination regimen used.

The main site of injury during small-bowel preservation has been thought to be the endothelium and basement membrane of the highly vascularized mucosa. The small intestine has a high concentration of the enzyme xanthine oxidase, rendering it liable to reperfusion injury. Free- radical damage has been implicated, particularly after prolonged preservation and after warm ischemic injury. Experimental preservation is possible for up to 48 hours in the rat and dog; the advantages of any one preservation solution are less obvious than with other organs. Sucrose, Collins', UW, and Bretschneider's solutions, and solutions with added dextran and containing free-radical scavengers have given more or less equivalent results. Pretransplant rinse has not been shown helpful in preventing reperfusion injury.

Polyethylene glycol has been suggested as an additive both to enhance preservation and to modify rejection in experimental small-bowel, heart, and pancreas transplantation. Small-bowel and small-bowel/liver transplantation have evolved from an experimental therapy to an accepted form of treatment for patients with intestinal failure and associated life-threatening complications. Factors that have contributed to improved patient and allograft survival include new immunosuppressive agents, better patient selection, and refinements in the surgical procedure. The upper limit of acceptability is currently 12 hours of cold ischemia.

Composite Tissues and Limbs

Reconstructive plastic surgery uses vascularized autografts of composite tissues to fill large defects. These grafts involve skin, subcutaneous fat, muscle, bone, nerves, and blood vessels. Severed limbs and digits can be replaced as autografts in complex operative procedures that can take many hours. Storage of the grafts relies principally on simple external cooling by refrigerated saline and on wrapping the grafts in cold saline-soaked packs during implantation. Vascular flushing has not been widely used for fear of damaging the small vessels requiring microvascular suture.

Flushing does not give any significant improvement over simple hypothermic storage in cold-preserving liquid. Recent studies on the cold preservation of skeletal muscle demonstrated that HTK was an effective preservation medium when used as bathing solution, whereas no clear advantage of vascular perfusion could be shown. The addition of antioxidants trolox and deferione to the storage medium improved muscle function after cold storage. Hypothermic storage usually adequately covers the periods required in clinical practice — usually less than 12 hours. However, in a composite murine whole-limb replantation model, cooling to -1°C in the absence of ice formation provided better protection than that seen at 4°C.

The challenge with this technology is to maintain strict temperature control such that ice does not spontaneously nucleate following a small drift in the temperature of storage. In general, tolerance of the various tissues within limbs and digits to ischemia varies: muscle and nerve are more sensitive than skin and bone. Restoration of the circulation to reimplanted limbs or other composite grafts of large bulk may release a large bolus of potassium into the circulation and can induce fatal hyperkaliemic cardiac arrest. After extended storage, the contained blood or preservation solution should be rinsed out with warm plasma or balanced electrolyte solution before release of the clamps.

New Developments in Organ Preservation

Ischemic Preconditioning

Organ preconditioning is a process whereby a brief event, be it transient ischemia, oxidative stress, temperature change, or drug administration, bestows on an organ a temporary tolerance to further insults by the same or similar stressors. Ischemic preconditioning (IP) is the exposure of the organ to a brief period of ischemia (e.g., 5–10 minutes) and reperfusion that has been shown to effectively protect it against subsequent warm ischemia and cold storage injury. Murry et al. first recognized organ preconditioning: While trying to create a larger myocardial infarction in a dog, they subjected animals to several brief episodes of myocardial ischemia and reperfusion before a protracted ischemic injury. Surprisingly, they observed that the area of infarction in the stressed animals was dramatically reduced — by up to 75% — compared with animals not subjected to the antecedent brief ischemia.

In addition to limiting infarct size, IP reduced myocardial functional stunning and postischemic arrhythmias and accelerated the recovery of muscle function after ischemia. Similar pictures of ischemic protection have been confirmed in multiple organs across many species. Preliminary studies in humans confirmed the efficacy of IP in the liver. In animals, ischemic preconditioning of lungs before prolonged hypothermic storage is protective of both compliance and gas exchange. There is strong evidence that adenosine is a key mediator in IP through its stimulation of adenosine A_2 receptors, which initiates NO formation, and causes activation of protein kinase C, AMP-activated protein kinase, and p38 MAPK.

These intracellular signaling pathways, once activated, not only trigger increased tolerance of the hepatocytes, and endothelial cells in the case of the liver, against ischemic insults, but they also cause quiescent cells to enter the cell cycle and initiate a regenerative response. Induction of heat-shock proteins HSP70 and heme oxygenase 1 (HSP32) has been implicated in the mechanism of preconditioning. HSP induction can reduce the nuclear binding of proinflammatory transcription factors and increase the antioxidant capacity of cells. This may contribute to a reduction in the formation of TNF-α and an attenuated inflammatory response in preconditioned livers.

Carbon monoxide, a byproduct of heme oxygenase 1 activity, was shown to activate p38 MAPK as a key mechanism of carbon monoxide-mediated protection against IR injury. Thus a combination of factors may contribute to the reduced injury and the improved long-term survival in animals subjected to preconditioning, including the increased tolerance to ischemic injury and oxidant stress, a reduced inflammatory response, and enhanced regeneration.

Higher-temperature Storage

The advantages of oxygenated perfusion and its specific use in organs (mainly kidney and liver) have been discussed above. Recently, there has been increasing interest in these techniques, with the abandonment of the principle of hypothermia as the cornerstone of preservation in favor of attempting to replicate homeostasis by supplying adequate substrates at normothermia or mild hypothermia (10°C); hence the description of "*higher-temperature storage.*" This status has been brought on by the increasing inability to expand donor pools and the increasing use of previously considered marginal organs (e.g., steatotic livers and non-heart-beating organs), which suffer more damage from cold ischemic storage.

The introduction of UW improved the quality of the cold-preserved grafts and remains the gold standard of organ preservation. Further improvement of UW was attempted, but without clear clinical effect. It is doubtful that further significant improvements of organ-preservation solution will be possible as long as the solution is based on a static ischemic principle. In contrast to the static method of cold preservation, normothermic oxygenated perfusion (NOP) could be a major step forward if it were possible to maintain the physiologic metabolism of the graft extracorporeally during preservation. To stop the process of biodegradation as it takes place under cold static preservation, a graft needs substrates and must get rid of metabolites. NOP provides substrates and, as long as it is being conducted with simultaneous dialysis, also allows for disposal of metabolites.

The technique of isolated liver perfusion for evaluation of preservation methods, either normothermic or at low temperature, has been known for many years. The potential advantages of NOP storage versus cold and static cold-flush preservation are obvious. First, no cold-ischemic injury is inflicted. Second, it is possible to monitor liver viability during perfusion by bile production, transaminase release, and metabolic function before transplantation. The most convincing evidence of NOP is the success of preservation, proven by primary graft function after transplantation.

Moreover, NOP must be compared with UW preservation, as it is the standard against which all modifications in preservation must be assessed. There are few examples in which this has been achieved, the most interesting of which is from Neuhaus' group in Berlin, using a pig model. This group developed an advanced liver perfusion circuit in which the organ was placed in a waterlogged and sealed chamber. Oscillating pressure profiles imitating the intraabdominal pressures improved perfusion of peripheral lobules significantly.

Further, the introduction of simultaneous dialysis of the recirculating perfusate allows for regulation of pH and physiologic electrolyte concentrations. Water-soluble toxins and increased amino acids are removed. The researchers demonstrated that it was possible to maintain a liver on NOP (37°C), for 4–5 hours, using a perfusate of pig blood from fed pigs under the assumption that it contains all necessary substrates, and to transplant it successfully, producing results that were comparable to immediately transplanted livers.

The scientists went on to demonstrate that NOP was superior to cold UW preservation when it was preceded by 1 hour of warm ischemia, which has far-reaching implications for the use of non-heart-beating organs. Working almost simultaneously, Friend's group in Oxford tried to simplify the NOP circuit while further exploring its use in preserving nonheart-beating pig livers. This group succeeded in maintaining the liver's function for 48 hours on NOP in a nontransplant model. These studies have obvious implications for non-heart-beating and steatotic organs that are more sensitive to cold ischemic injury than cadaveric organs.

Gene Therapy in Organ Preservation

The ability to modulate the damage mechanisms following organ preservation by gene therapy has moved from theory to practice over the last 5 years. Specific pathways, such as the regulatory genes controlling apoptosis (e.g., Bcl-2) have been shown to be amenable to transfection and to reduce hypoxic cell damage. Improved survival after liver cold preservation was demonstrated by transfection of Bag-1 (another of the proteins that interacts with BCL-2). There was also a possible effect on the heme oxygenase pathway (a member of the stress protein family), and this supported previous work in which transfection of HO-1 was also protective in experimental liver storage and transplantation.

Combined transfer of genes for antioxidant enzyme super oxide dismutase and cytoprotective nitric oxide synthase provided enhanced function in a model of cardiac preservation and transplantation. There remain many hurdles to overcome before such techniques reach clinical application, including the nature of the transfection agent and the timing of gene induction, but these will undoubtedly become the center of attention in the near future.

Freezing and Vitrification

The biophysical events dictating survival of cells after exposure to very low temperatures, ice crystal growth, and the need for protections by cryoprotectants have been dealt with in detail elsewhere in this book. Attempts at cryopreserving organs have a long history, but occasional reports of success have proved difficult to repeat in a consistent fashion. For example, short-term freezing of kidneys causes very substantial damage, and such organs never regain normal function.

Thus, unfortunately, freezing of organs, so attractive in concept, in establishing organ banks by true suspended animation and indefinite storage, has not proved feasible to date. There are many characteristics specific to organs that dictate this poor survival, but it is worth commenting on a few here.

Although it is relatively simple to add and remove cryoprotectant mixtures from cell suspensions by mixing solutions, the ability to achieve this in a large, three-dimensional structure like a mammalian organ can only be achieved by controlled vascular perfusion. Because osmotic damage can be induced by exposure to high concentrations of such chemicals, perfusion must be allied with integrated control of cryoprotectant concentration change. This has been achieved in experimental studies but has not so far been applied in the clinical setting.

For single cells, the distribution or orientation of extracellular ice crystals is largely irrelevant, but in organs in which cells are packed in high density, with important connections to the basement membrane or other cells, growth of ice can cause significant structural damage. Given that organs are of large size, with a low surface area–to-volume ratio, manipulation of cooling and warming rates are restricted. Because slow cooling rates are compatible with the survival of many mammalian cells, ice is not a problem during the cooling phase, but it may have a significant negative effect during warming, where rapid rates are usually imperative to avoid damaging events linked to ice recrystallization.

Application of electromagnetic warming has been suggested for many years, but significant biophysical and equipment-related problems remain to be sorted out, although progress is being made in this area. The cellular heterogeneity of organs such as the kidney is another problem, because different cell types vary in their requirements for successful cryopreservation. The vascular system appears especially vulnerable, and the vascular endothelium cells are disrupted by freezing, leading to occlusion after revascularization.

Cooling to subzero temperatures without freezing (supercooling or, more correctly, undercooling) diminishes problems caused by ice formation, but a depression of only a few degrees in storage temperature can be achieved, and the toxicity of cryoprotectants is still excessive. An alternative method of cryopreservation is based on total vitrification of the organ. In this approach, the organ or tissue is first perfused with a high concentration of cryoprotectants so that on cooling, neither intracellular nor extracellular freezing occurs. The solution vitrifies (i.e., solidifies into a glassy state) at temperatures below approximately −100°C, marking the effective end of biological time. Extremely high concentrations of cryoprotectants are needed for this approach. Again, problems of addition and removal of these agents and control of cooling and warming remain, but investigations are continuing, with reported improvements in some aspects of the technology.

9

Specific Biopreservation

The application of low temperatures to select anatomical regions is of well-established clinical benefit when the process is appropriately controlled. Moderate cooling of a tissue followed by warming over short intervals (minutes) causes reversible effects on the cellular physiological processes, including perturbation of electrical events, while with deeper and extended cooling, ice formation occurs resulting in site-specific lethal consequences. The employment of low temperature for clinical and research use includes, but is not limited to, applications of thermal therapy: (1) hypothermic storage of cells, tissues, and organs; (2) cryotherapeutic treatment for the selective ablation of cells and tissues; and (3) cryoangiogenic-based tissue repair. Although these applications have certain distinctions, the underlying principles of cell and tissue responses to low temperature provide the foundation or linkage between each of the respective applications. Current expansion in the field of cryotherapy stems from the growing demand for multidisciplinary studies in which knowledge from one specialty area may be transformed to benefit other specialty areas.

The therapeutic uses of cold temperatures had their origin in ancient times and were diverse in scope, ranging from local applications to general body hypothermia. Local application of cooling agents were used for the treatment of open wounds and other forms of trauma, hemorrhage, fever, and a variety of clinical disorders, including infected ulcerating cancers of the breast. Total body cooling by immersion in ice water was used for intractable fevers, mental derangements, and other diverse diseases. Modern cold therapy has similarities to the practices of antiquity. In current times, the body temperature is

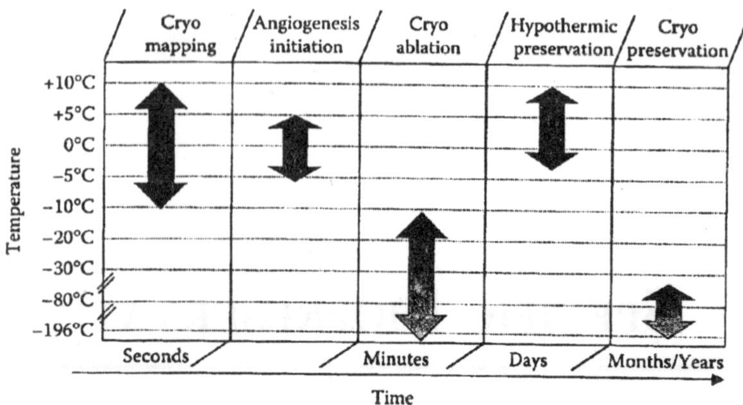

Fig. 9.1. Illustration of the thermal zones and temporal duration for various clinical applications of cold.

commonly lowered in some types of cardiac surgery, in some difficult cancer operations, and in the management of stroke in order to reduce the oxygen consumption of the brain. Today also, cold applications are used commonly after soft tissue trauma to relieve pain and help prevent swelling of the injured part. These diverse uses of cold can be considered cryotherapy, but none uses the freezing of tissue to produce the desired therapeutic result. This chapter deals with the use of freezing temperatures in therapy.

Development of Cryosurgical Techniques

Local tissue freezing was first described in the mid-1850s when Arnott, a physician in England, used salt solutions containing crushed ice at temperatures in the range of -18 to -24°C to freeze cancers in accessible sites such as the breast and uterine cervix. He wrote that "congelation arresting the accompanying inflammation, and destroying the vitality of the cancer cell, is not only calculated to prolong life for a great period, but may, not improbably, in the early stages of the disease, exert a curative action." The treatment with the frigid fluid was given with an irrigation device that Arnott had developed. The device could be used only on accessible tissues and had limited freezing capability, but the benefits in patients with cancer included relief of pain, reduction in the size of the tumor, and lessening of bleeding and discharge. The amelioration of pain by this technique received noteworthy attention by physicians of that era.

Late in the 1 800s, the so-called permanent gases (oxygen, nitrogen, hydrogen) were liquefied and in 1892, Dewar developed the vacuum flask to hold them. As the century ended, liquid air and solidified carbon dioxide were available for therapeutic use on localized lesions of the skin. In the next six decades, the treatment by freezing was limited to accessible lesions, such as those in the skin or uterine cervix. The cryogen in common use was solidified carbon dioxide, applied either directly or in devices developed for its use. The word cryotherapy was coined midway through these years in a publication that described the use of solidified carbon dioxide in a new device. During those years, the use of freezing techniques to treat disease of deep tissue or visceral disease received little attention. The notable exception was the investigation of Temple Fay, a neurosurgeon in Philadelphia, who used local and general refrigeration techniques to treat cancers. Fay implanted metal capsules, connected to an irrigation system using ice water or saline solution, circulated by gravity, in tumors of the brain. The tissue response to the freezing featured reduction in tumor size and amelioration of symptoms, just as described by Arnott nearly 90 years earlier.

The era of modern cryosurgery was preceded by experiments in the 1940s and 1950s in which researchers produced focal tissue lesions by freezing to gain information about the pathophysiology of disease. Hass and Taylor, using an apparatus cooled by the use of pressurized carbon dioxide, produced discrete necrotic lesions, 2 to 3 cm in diameter, in the brain, the heart, the liver, and the kidney. Other investigators in these years, using diverse types of apparatus, showed that necrotic lesions could be produced in diverse organs and demonstrated the safety of the freezing procedures.

The modern era of cryosurgery began with the introduction of an automated apparatus cooled by liquid nitrogen (–196°C) by Cooper and Lee in 1961. Originally, the apparatus was intended for the treatment of brain disease, especially Parkinsonism. However, the apparatus was redesigned and widely used for many types of neoplasms in the 1970s. Then, after clinical trials defined appropriate usage, cryosurgery became a well-established method of ablating benign and malignant lesions, especially those tumors that were a problem in treatment by conventional methods. Modern advances have refined cryosurgical procedures, broadened their applicability, and enhanced their reliability. Percutaneous application of cryosurgery in the treatment of localized cancer is considerably less invasive than conventional surgery and is free of the side effects associated with chemotherapy or radiation. Success in reducing morbidity through percutaneous cryosurgical procedures has promoted further technical advances in the instrumentation available to employ cryosurgical methods. Technology has enabled the development of precise instrumentation through which effective cryogens (e.g., liquid nitrogen, nitrous oxide, argon, etc.) can be circulated, providing the means to deliver extremely low temperatures to tissues located internally as well as near the body's surface. In the 1990s, technological advances included newer types of apparatus, using pressurized argon (–185.7°C) or nitrous oxide (–89.5°C) in devices which cooled probes by the Joule-Thomson effect, that is, passing the gas through a restrictive orifice into the heat exchange chamber of the probe, as well as devices that circulated super-cooled LN_2 (–210°C). The development of intraoperative ultrasound to monitor the freezing procedure, and the availability of minimally invasive operative methods, has permitted wider and new uses of cryosurgical techniques.

During these years, an improved understanding of the mechanism of injury to cells and tissues has been achieved. Earlier investigations on the tissue injury due to frostbite formed a good foundation to the understanding of the nature of the injury *in vivo*. Cryobiological investigations on the preservation of cells by freezing contributed to the knowledge needed to apply freezing to the destruction of cells These investigations provided a good foundation upon which to build cryosurgical techniques. The variable response of cells and tissues to freezing injury was recognized. The importance of careful attention to the elements of the freeze-thaw cycle, including tissue temperature, duration of exposure, and slow thawing, was recognized as critical to the ablation of tumors. However, in recent applications to non-neoplastic disease, the complex balance between preservation and destruction was also recognized. This balance is being tested by the expanding uses of cryosurgery and by the need to improve the results of cryoablative techniques. The new frontier is the development of techniques to manipulate the response of cells at low freezing temperatures in a way that would selectively preserve or destroy tissue, influence the process of healing, and increase the efficacy of therapeutic freezing techniques.

Tissue Response to Freezing Injury

The tissue response to injury by freezing is dependent upon many factors and is determined only in part by the rate of cooling or lowest temperature produced in the tissue. These physical factors are important, but other factors, such as the type of tissue and the vascular reaction to freezing temperatures, are significant factors in determining the fate of tissue frozen *in vivo*. With this complexity of critical elements of injury, variations in the response of cells and tissues occur, ranging from inflammatory to destructive. Minor freezing of tissue produces an inflammatory response, featuring the infiltration of leukocytes and development of edema. For example, the treatment of retinal detachment requires the use of warm freezing temperatures to produce an inflammatory response that promotes adhesion of the detached tissue. Deep-freezing techniques cause destruction of tissue and are used in the treatment of tumors.

The causes of tissue injury from freezing generally have been attributed to two major mechanisms, which are the direct injury to cells due to ice crystal formation and the microcirculatory failure that occurs in the thawing period. In freezing, the temperature produced in the tissue is of prime importance in determining the tissue response. A short exposure of tissue to temperatures in the range of 0 to -10°C will produce a relatively minor injury, featuring inflammatory cell infiltration, new blood vessel formation, and collagen production. Since tissues can supercool to perhaps -15°C, little or no ice crystal formation occurs, so cell death is minimal. On the other hand, colder temperatures produce ice crystal formation, well known to be initially extracellular, but as the temperature falls, intracellular ice formation occurs, and the cells die in progressively greater numbers from mechanical forces or biochemical effects. Practically certain destruction of cells occurs at temperatures colder than -40 to -50°C, commonly termed a "*lethal temperature*" in cryosurgical practice.

Recent research has shown that apoptosis is a mechanism of cell death after cold injury. Apoptosis, or gene regulated cell death, has been recognized as a mode of cell death in normal tissues and in diverse pathologic conditions, including heat injury, cancers, cytotoxic chemotherapy, and irradiation. Apoptotic cells are characterized by non-random DNA cleavage, blebbing of membranes, phospholipid inversion in the membranes, and irradiation. Several recent investigations have shown that apoptosis is an important mechanism of cell death in prostate cancer cell lines when the temperature did not fall sufficiently low to kill all cells immediately. Sublethal freezing temperatures have been shown to induce apoptosis in colon cancer cells *in vitro* also. Most investigations of apoptosis caused by freezing have been *in vitro*. Investigations *in vivo* are few in number. However, Steinbach et al., producing cryogenic tissue damage in the cerebral cortex of mice, described apoptosis cells in the periphery of the previously frozen tissue. Romaneehsen et al., freezing and thawing a non-small-cell cancer implanted in nude mice, also described apoptotic cell death. Since apoptosis occurs in so many circumstances, it is reasonable to expect apoptosis to occur in cryogenic lesions *in vivo*. Indeed, Baust and Gage recently reviewed the area of the molecular basis of cryosurgery and described the critical role of apoptosis and necrosis in cryotherapeutic outcome.

When cryobiological research is performed *in vivo*, another important mechanism of cell death comes into play, that is, the effect of freezing on the blood supply to the tissue. The initial response to the cooling of tissue is vaso-constriction and a decrease in blood flow. With freezing, the circulation ceases. Experiments *in vitro*, using liver and breast tissue, have shown that ice crystals form in the blood vessels, and then preferentially propagate through the vascular system where no cell membranes form a barrier. When the tissue thaws, the circulation returns, and the tissue appears to be hyperemic for a brief interval. However, endothelial cell damage leads to increased permeability of capillaries, edema, platelet aggregation, and thrombosis, so the outcome is vascular stasis in 30 to 60 minutes. These changes have been shown to occur at temperatures as warm as -20°C. The loss of blood supply deprives the cells of any chance for survival. The injury to tissue from vascular stasis complicates cryosurgical research because this mechanism is not operative in research *in vitro*.

Cryogenic Lesion

The type of lesion produced by freezing depends on the severity of the trauma. Minor freezing injury, as would be produced by rather brief exposure to temperatures warmer than about -10°C, causes an inflammatory response, featuring hyperemia, edema, and the infiltration of neutrophils, monocytes, etc. Deep freezing, as would be produced by tissue temperatures colder than -20 to -30°C for several minutes, produces a sharply circumscribed necrosis that closely corresponds to the volume of tissue frozen. Between these two extremes, perhaps especially in the temperature range of -10 to -20°C, the potential for special therapeutic applications permitting selective destruction may be found.

The cryogenic lesion associated with deep freezing, as is required in the treatment of tumors, is a volume of coagulation necrosis. After freezing and thawing, initially the tissue is hyperemic or congested and edema forms. The extent of injury is not obvious, except on electron microscopic studies, which show disruption of cellular components and their membrane. In about two days, the extent of the injury becomes visually evident. The necrotic tissue will be resorbed or will slough, depending upon its location. For example, skin lesions will slough, but hepatic lesions will be resorbed. Tissue with a large amount of fibrous stroma resists change in structure, so the process of sloughing is prolonged, perhaps to several weeks. Major blood vessels, nerves, and cartilage resist structural change, even though devitalized by freezing. After freezing and thawing, bone is unchanged in appearance and maintains form and function, even though the devitalized bone is weakened by the process of slow resorption and replacement.

The central part of the cryogenic lesion, that tissue which was closest to the cryosurgical probe or heat sink, is practically always totally necrotic. At the periphery of the previously frozen volume of tissue, which is the region where cooling was very slow, the temperature rather mild (0 to –20°C), and duration of freezing relatively short, the cells have a chance of survival. In this border area, some cells live, some cells are dead, some are apoptotic, and some are in the balance between life and death. This border area is commonly quite large in comparison to the volume of tissue that was frozen to temperatures 40°C and colder.

Cell and Tissue Responses to Cryosurgery

The manner in which the cryogen wound heals depends upon the severity of injury and the type of matrix of the tissue. Minor freezing injuries commonly heal without tissue loss. In severe freezing injuries, the process of healing depends on the type of matrix. Some examples may be cited to demonstrate how the matrix of different tissues affects healing. In the skin, the resistance of collagen fibers and fibroblasts to damage from freezing is responsible for the favorable healing. Damaged collagen is absorbed and slowly replaced by new collagen during the days of healing. Hair follicles, glands of the skin, and melanocytes are sensitive to freezing injury and these elements do not regenerate after exposure to temperatures in the –10 to –20°C range. Nerves are also sensitive to cold injury. Temporary desensitization of the nerve occurs in the 0 to –5°C range, but function returns when the tissue is warmed. Long-term loss of function is produced at colder temperatures that range from –15 to –20°C. The axons and Schwann cells of the nerve degenerate, but if the nerve sheath is not disrupted, function of sensory and motor nerves is likely to return, though this restoration may take many months. The structurally intact perineurium, though devitalized by freezing, serves as a conduit for regrowth of axons and eventual restoration of function. Muscle cells are sensitive to freezing injury as well.

In contrast to the microcirculation, which is thrombosed after freezing, the function of major blood vessels is preserved following cryotherapeutic treatment. Though the blood vessel wall is devitalized and the endothelium is lost, the structure of the blood vessel remains intact. Generally, only slight fragmentation of the collageneous and elastic fibers are seen. The blood pressure maintains the lumen, so function as a conduit continues. The process of repair features smooth muscle proliferation, collagen and elastic fiber repair, and endothelial restoration, resulting in some degree of intimal hyperplasis, narrowing the lumen to some extent.

Bone resists structural change after freezing, though the osteocytes are very sensitive to freezing and are destroyed at about −10°C. Bone, devitalized by freezing, is slowly resorbed and replaced by new bone, a process that may require months, the time depending upon the volume of bone frozen. The resorptive process weakens the strength of the bone, so fracture may result from even normal stress.

Cellular tissue, including most organs and mucosa of the oral cavity and in other sites, lacks the strong collagenous structure of skin and has little fibrous tissue to maintain structure during healing. Therefore, healing is generally from the wound edges, resulting in some scarring, but not as much as produced by hot thermal injuries.

Clinical Implications

Cryotherapy has a wide range of applications, including anesthetic, preservative, and destructive uses. Relief of pain by the application of cold solutions or compresses is the oldest method of anesthesia. The manner of use was described in ancient literature, and later by Larrey to aid in the amputation of limbs of soldiers in Napoleon's armies. Further interest in cold anesthesia was caused by Arnott's description of its impact in diverse conditions in 1850. Bird called Arnott "a pioneer in refrigeration anesthesia" in an interesting report. The advent of general anesthesia in the mid-1850s and subsequently of local anesthetic agents about the same time sharply reduced the need for refrigeration anesthesia, but today cold agents are in wide use for the local surface treatment of soft tissue trauma, and apparatus is available for the treatment of pain by the freezing of the appropriate nerves. This introduces the subject of selective or function-preserving cryotherapy.

Selective Cryotherapy

The tissue response to a range of temperatures provides the basis for choices in therapy. The warmer freezing temperatures would be an appropriate choice when selective cell responses are the goal. When tissue destruction is the goal, then very cold temperatures are required. *In vitro* experiments have shown that cells can survive at very low temperatures. A wealth of information is available on this subject. For example, several reports may be cited. Zacarian long ago showed that cancer cells in suspension could survive in small numbers after exposure to temperatures colder than -60°C. More recently, Bischof et al., using rat prostate tumor slices, found that some cells could survive after freezing to a temperature of -40°C at a rate greater than -10°C/minute. Hollister et al. also described occasional cell survival, as measured by ability to grow in culture, of cells frozen to -60°C or colder. Further, Clarke et al., Baust et al., and Znati et al. have demonstrated that in select *in vitro* models, cancer cells can survive exposure to temperatures approaching -100°C. The same results are not likely to be seen in work *in vivo*, but the resistance of cells to freezing injury in these experiments is impressive.

Selective destruction of cells by freezing temperatures has considerable therapeutic potential, which is based on the differences in the sensitivities of different cell types to cold. These variations are best identified in single short freezing cycles because longer duration or repeated cycles destroy most cells. Experiments *in vivo* show that the temperatures at which differences in cell susceptibility are evident are in the 0 to -3 0°C range. Therefore, in freezing therapy, the possibilities of selective preservation of some cells, while destroying other cells, can be explored. Furthermore, selective functional compromise with later return of function has applications in cryotherapy. Related to this thought is the possibility that a nondestructive response may induce changes in the healing process. These uses of cryotherapy commonly require modest freezing temperatures, that is, in the 0 to -30°C range.

The potential applicability of selective destruction/preservation has not yet been fully investigated. One example may be cited. Exploiting the cryosensitivity of melanocytes, freezing is commonly used in the treatment of lentigines, which are pigmented lesions of the skin, usually benign, sometimes malignant. The healing of the wound is favorable because the keratinocytes, fibroblasts, and collagen matrix are not seriously injured at tissue temperatures warmer than -30°C.

Cryomapping

An example of the use of freezing to produce functional compromise is found in the use of cold temperatures to inhibit the function of nerves. As early as the 1960s, Cooper used temperatures slightly cooler than 0°C to temporarily desensitize nervous tissue in the brain in an effort to determine that the cryosurgical probe was located at the appropriate site. Then, once the probe was in the correct site, the probe was further cooled to produce the intended therapeutic effect. The same principle is commonly used today in the surgical treatment of disease of the conduction system of the heart.

Commonly termed "*cryomapping*," the procedure requires that a flexible steerable catheter probe be placed in the heart in a position at the suspected problem in impulse conduction or irritable focus of excitation. Then the probe is cooled to produce tissue temperatures in the 0 to –5°C range, which is sufficient to temporarily interrupt nerve function. If this site is not the problem, the probe is warmed. However, if the site is the problem, the faulty conduction/excitation returns on thawing, confirming that the diseased site has been identified. The treatment requires that the probe be cooled to –70 to –80°C for the time needed to ablate the tissue, generally about two or three minutes. The temperature at the site of the abnormal conductive tissue is in the range of –20 to –30°C.

While these studies have reported on the successful clinical application of this therapy, an understanding of the mechanisms of action, responses to low-temperature therapeutic intervention at the cellular level, as well as the window of therapeutically relevant temperatures in cardiovascular systems remains unclear. Recently, a study by Snyder et al. investigated the cellular and molecular responses of cardiac systems to thermal excursions utilizing an *in vitro* neonatal cardiomyocyte model. The purpose of this study was to identify and define thermal zones of physiological reversibility (cryomapping) and cellular destruction (cryoablation) in cardiac cell types in an attempt to further expand the knowledge base on cellular responses to cryotherapy in order to help guide clinical technique and device development.

This study demonstrated that the application of cryotherapy in cardiac systems provided an effective and more controllable alternative to heat-based therapies such as radiofrequency. Additionally, the study served as a useful procedural guide for the application of low temperature as well as suggesting a mode of treatment in sensitive areas of the heart, where tolerance to mapping and ablative power are equally important, such as near the atrioventricular node, sino-atrial node, and pulmonary veins.

Cryoangiogenesis

Attempts to alter the healing process, eliciting specific responses in terms of angiogenesis or myogenesis, are at the forefront of current research, especially in relation to the heart. All tissues require vasculature access for nutrients, oxygen, and waste removal. Vasculogenesis during embryonic development provides the basic vasculature structure, which is expanded by the process of angiogenesis. Endothelial cells proliferate and migrate to form microtubules, followed by the modification of the extracellular matrix and finally the formation of patent vessels. Adult angiogenesis begins with angiogenic sprouting initiated by vasodilatation, followed by the destabilization of the existing vessel to allow for the migration of the proliferation cells and the eventual maturation. However, angiogenesis is typically associated with embryonic and neonatal vascular development. In adults, angiogenesis is typically associated with vascular and tissue repair, as well as in the formation of nutrient supplies for cancerous tumors. Stimulation of angiogenesis by exposure to nitric oxide or hypoxia has been shown *in vitro* and *in vivo.*

Cryogenic injury in tissues causes an angiogenic response as part of the reparative process. New blood vessels enter the necrotic tissue at its border with viable tissue, and repair progresses as the vasculature is reestablished in the lesion. Recently an effort has been made to provide a new blood supply to the ischemic myocardium by using a thin cryoprobe to penetrate the ischemic area many times, hoping that the freezing injury will induce angiogenesis and create new channels for perfusion of blood. A measure of doubt regarding the potential of these experiments is reasonable, especially since laser-created channels do not remain patent.

However, the use of thermal therapies to enhance angiogenesis in myocardial systems has shown positive results, including vascular growth and induction of angiogenesis molecular signaling molecules, such as vascular endothelial growth factor (VEGF). One recently identified potential area of application of the angiogenic process is the use of mild cryotherapy to induce angiogenesis in ischemic tissue. This approach is designed to take advantage of the myriad cellular and molecular responses to low-temperature insult to stimulate cellular proliferative activity and revascularization. This is a new area of ongoing research and initial success has recently been reported in an *ex vivo* model by Snyder et al. In this study, the effect of low temperature on the activation of the angiogenic process in an aortic ring model was investigated. Analysis of cellular outgrowth (smooth, muscle) revealed that there was in fact an activation of the angiogenic process by low-temperature exposure. The data demonstrated that this response could be modulated by both the thermal and temporal components of tissue exposure.

Short intervals of 30 seconds required subzero temperature exposure to elicit response. The greatest response of 30 seconds of exposure was observed in the –5.5°C condition. This exposure resulted in a high degree of cell outgrowth and microvascular formation. Extension of the exposure interval to 60 seconds resulted in a change in the response of the Rat Aortic Ring Segment (RAR). Following 60 seconds of –6.6°C thermal exposure, RAR illustrated a significant reduction in the level of outgrowth compared to controls. In the 60-second conditions, samples exposed at 5°C yielded the greatest degree of outgrowth and organization. In addition, the response of the –2.5°C exposed samples remained relatively unchanged between the 30-second and 60-second exposure intervals. While this area of investigation remains in the early stages, nonetheless it is providing new avenues for the interventional therapeutic initiation of neovasculariztion in target tissues.

Cryoangioplasty

With regard to myogenesis, interest was stimulated by the incidence of restenosis after coronary angioplasty, which is about 30% over a year or two. Balloon dilation of the stenosed artery is associated with disruption of the endothelium and underlying tissue, and this injury is followed by a reparative response featuring inflammatory cell infiltration and proliferation of smooth muscle cells. This response produces initial hyperplasia, narrowing the lumen, and perhaps causing restenosis. Also, angioplasty-induced damage along with increased vessel permeability triggers a thrombogenic response that in turn serves as a strong mitogenic and proliferative signal for medial smooth muscle cells. This signaling elicits intense smooth muscle cell proliferation into the vessel lumen, consequently constricting the vessel lumen and greatly hindering blood flow.

Furthermore, it has been shown that altered fluid mechanical shear stresses in the sclerosed vessel segment can have a drastic influence over both vascular endothelial and smooth muscle cell structure and function. In view of these details, it is important to encourage a rapid recovery of the vascular endothelium while re-closure due to the overgrowth of medial smooth muscle cells into the lumen is discouraged. Accordingly, an optimized cryotherapeutic option may rely on the application of "*preservative*" temperatures for endothelial cells and ablative temperatures for smooth muscle cells.

Efforts to improve the efficacy of vascular therapy have focused to a great extent on the stimulants and inhibitors of smooth muscle cells migration, proliferation, and phenotype. Today, commonly used lipid-lowering drugs, such as simvastatin and gemfibrozil, have also been shown to increase levels of TGF- 1 (a suppressor of inflammatory processes and proliferation of smooth muscle cells) and inhibit the release of platelet-derived growth factor-B (PDGF-B — a powerful promoter of smooth muscle cells migration and division) in the neointimal. Using *in vitro* and *ex vivo* models alike, investigators have demonstrated the potential therapeutic value in blocking myointimal hyperplasia following medical intervention (angioplasty or vascular graft) by blocking either the PDGF growth factor or smooth muscle cells PDGF receptors themselves.

Some of the latest efforts to curtail smooth muscle cells hyperplasia post-intervention have entailed the placement of drug-eluting stents. However, the presence of a stent has been known to further complicate the matter by adding inflammatory effects due to an immunological response to the foreign mesh. Accordingly, freezing temperatures are being used experimentally in an effort to control smooth muscle proliferation.

Cryogenic injury to major blood vessels, including the coronary arteries, causes some intimal hyperplasia, though this reaction tends to regress over several months. The experimentation to control the smooth muscle proliferation, which follows angioplasty, features the use of a catheter cryoprobe into the injured vessel, then freezing the stenotic area to a preselected temperature. If benefit is to be achieved, the tissue temperature will have to be in the –10 to –15°C range. Experimental work by Hollister et al. supports this temperature range as an appropriate goal in therapy. Their work subjected cultured human coronary artery endothelial (CAEC) and smooth muscle cells to a range of hypothermic and freezing temperatures. They showed that smooth muscle cells (CASMC) were more sensitive to cold injury than endothelial cells in the range of 0 to –15°C.

The endothelial cells had a better proliferative capacity than the smooth muscle cells. The work suggests that it might be possible to apply a cooling temperature that would preserve the endothelial cells and inhibit smooth muscle proliferation. The lasting susceptibility of smooth muscle cells to the measures applied in this study suggests the potential for endovascular application of cold to act as a "*virtual stent*," allowing the endothelium to recover, thereby minimizing neointima formation. However, Cheema et al., in *in vivo* experiments using rabbits, comparing intravascular cryotherapy after iliac artery balloon angioplasty with balloon angioplasty alone, found no evidence of benefit from cryotherapy.

In this study, the average arterial wall temperature was -26°C, which is too cold to permit benefit in terms of a selective cell response. At present, the temperature, which will provide a selective response in muscle and endothelial cells *in vivo*, is not known. Future experiments should include, as adjuncts to freezing, the use of pharmaceutical agents or other methods to inhibit smooth muscle proliferation and promote endothelial growth.

The reason(s) for the marked differential in sensitivity exhibited by smooth muscle versus endothelial cells remain(s) unclear, but may not be entirely unexpected. Variable success in the application of freezing has been reported dependent on the tissue type being treated. The differences among tissue types have historically been attributed to changing thermal conductivity with varying degrees of blood perfusion or inherent water content differences in tissue. More recently, our experiences applying various cold exposure regimes *in vitro* have yielded varying degrees of cell type-specific sensitivity to cold. These studies have ranged in scope from hypothermic storage to cryopreservation, to cryoablative methods and have explored inherent cellular/molecular responses to low-temperature exposure without the contribution of extraneous systemic factors, such as thermoconductivity and perfusion-related issues.

An alternate explanation for differential sensitivity may lie in the varying threshold for the activation of programmed cell death or apoptosis. Not only does apoptosis play a significant role in the progression of atherosclerosis, but it also appears to be involved following anoxia-reoxygenation of allografts, as well as freeze response of cells during and following cold exposure. Increases in the apoptotic protein Fas in rat aortic transplants have been identified in the developing lesions of allograft vasculopathy.

Anoxia-reoxygenation of human CAEC has also been shown to decrease levels of calcium-dependent nitric oxide synthase (cNOS) and anti-apoptotic Bcl-2 while increasing the Fas receptor. As mentioned previously, apoptosis has also been linked to freezing. A post-freeze apoptotic contribution to cell death has been identified in human prostate cancer cells and colon carcinoma cells. Further, studies from our laboratory into the mode(s) of cell death by which CAEC and CASMC are succumbing to cold exposure have revealed elevated caspase activity in both cell types, suggesting a possible apoptotic contribution in this system as well.

Recent studies by our group have demonstrated an apoptotic contribution to cell death in both CAEC and CASMC following a transient mild freeze exposure. The sustained apoptotic DNA laddering and procaspase-9 cleavage suggest a greater role of programmed cell death in the cold-sensitive CASMC. Taken as a whole, these data begin to provide a molecular basis for the endovascular application of cryotherapy for the post-operative alleviation of restenosis.

Cryosurgical Technologies and Approaches

Cryosurgical techniques directed at tissue destruction, as is required in the treatment of tumors, requires the use of apparatus and a cryogenic agent to produce tissue temperatures colder than -40°C. Citing their normal boiling points at atmospheric pressure, liquid nitrogen (-195.8°C), argon (-185°C), and nitrous oxide (-89.5°C) are the cryogens in common use today. Liquid nitrogen cools by a change in phase, that is, liquid to gas. Argon and nitrous oxide are used in pressurized cylinders and cool by the Joule-Thomson effect.

Liquid nitrogen is the coldest agent in use, but the need for insulation of the feed conduit limits the ability to construct probes smaller than 2–3 mm in diameter. Pressurized nitrous oxide or argon, used in Joule-Thomson-type apparatus, will not produce as low a temperature as liquid nitrogen; therefore, in use of tumor therapy, a greater number of probes are required in comparison to liquid nitrogen. The Joule-Thomson apparatus does not require insulation of the cryogen feed conduit, so probes as small as flexible catheters for intravascular insertion and use in the treatment of cardiac tachyarrythmias can be constructed.

Cryogens are applied to the treatment area by direct application (spray or pouring) and by use in closed systems (a cryosurgical probe). The spray technique, almost always with liquid nitrogen, is commonly used in dermatological practice for the treatment of skin lesions. Liquid nitrogen also can be poured onto the tissue, typically into the cavity formed by curettement of a bone tumor. Liquid nitrogen, argon, and nitrous oxide are used in probe systems with heat exchange surfaces for application to the diseased tissue. The capacity of the probes to freeze depends upon the choice of the cryogen, the temperature of the probe, the area of contact with the cold probe, and the heat brought to the area by the blood supply of the tissue. The wide choice of cryogens and apparatus provides the opportunity to select a technique suited for the disease.

Cryoablative techniques must be done in a manner that will destroy an appropriate volume of tissue, a goal that is of special importance in the treatment of cancers. In general, cryosurgical treatment of tumors is quite successful and has the advantage of commonly being a minimally invasive technique. Good results have been achieved in the cryosurgical treatment of diverse tumors, including those of the skin, liver, kidney, prostate, bone, bronchus, and other sites. The technique offers the possibility of curative or palliative benefit in some difficult problems in the management of cancer.

On the other hand, the goal of a cure of cancer is not always achieved. Persistent disease in the cryo-treated sites, for example, is commonly about 20–30% for liver tumors and about 10% for prostatic cancer. Persistent disease is a sign of failure to properly judge the extent of the disease or a failure in the technique of freezing. In this article, only the latter can be considered. Fortunately, progress toward optimization of cryosurgery is continually being made.

Cryotherapy of tumors requires that a cryosurgical probe be applied to the tumor, generally by insertion into the tissue. The probe is used as cold as attainable in order to freeze the tissue as rapidly as possible, and so-called lethal temperature is produced in the tissue. The tissue is allowed to thaw spontaneously, and then the cycle is repeated. The tissues close to the probe freeze rather rapidly and attain the coldest temperature, but the isotherms extend radially, so even a short distance (1 cm) from the probe, the freezing rate is considerably slower and the tissue temperature is warmer. Whatever the volume of the frozen tissue, the border is about 0°C. The –40°C isotherm, which is a commonly accepted therapeutic goal, is only a small percentage (about 30%) of the frozen volume.

Repetition of the freeze-thaw cycle brings the lethal isotherms close to –20°C, but this still leaves a periphery where the cells were exposed to temperatures between 0° and –20°C. The cells in this region may be viable. To produce a tissue temperature of –40°C at a safe distance outside the limits of a cancer requires a very aggressive freezing technique, which is not easy to achieve and not too practical in some sites, such as the prostate gland. The proximity of the rectum and the complication of rectal fistula limits aggressive freezing in this site. For these reasons, cryosurgery needs adjunctive therapy to increase the lethal effect in the periphery of the cryogenic lesion.

Adjunctive Cryotherapy

Adjunctive therapy may consist of cytotoxic drugs, irradiation, antifreeze proteins, apoptotic promoters, or other agents. The most commonly used adjunctive therapy is a cancer chemotherapeutic agent, and the use of the drug after cryosurgery has been thought to improve survival of patients. However, a measure of caution in the use of the drug is needed. The coagulopathy or complications due to the considerable cytokine release, which follows large-volume freezing, especially of the liver, may be unfavorably affected by the administration of the cytotoxic drug.

Nevertheless, advantage from the use of adjunctive chemotherapy is likely. Cooper et al., in animal experiments, concluded that the administration of cyclophosphamide potentiated cell-mediated immunity. Clarke et al., freezing prostate cancer cells *in vitro,* demonstrated that a combination of freezing and 5-fluouracil was associated with a greater reduction in cell survival than either freezing or the drug alone. Mir and Rubinsky, freezing melanoma cells *in vitro*, and exposing them to bleomycin, reported that the cold injury made the cells permeable to the drug, which is of therapeutic interest because bleomycin does not normally enter the cells.

The optimal dose and time of delivery of cancer chemotherapeutic agents is not established. The thought that the drugs can be concentrated in the tumor area, as first suggested by Benson, is an interesting and perhaps practical approach. The rationale is based on the failure of the microcirculation shortly after the tissue thaws. Drugs given systemically during the thawing period would become locked into the tumor area, unable to gain access to the general circulation, as vascular stasis develops. Homasson et al., using bleomycin, has shown that this drug sequestration occurs in bronchial cancers. The same principle would apply to drugs injected into the thawed tissue. Such drugs do not gain access to the general circulation.

With other adjunctive agents, there is little or no demonstration of efficacy in clinical trials. Cooled cells have shown increased radio sensitivity. Though the role of irradiation as an adjunct to cryosurgery is not clear, the use of this agent should have a deleterious effect on cells in marginal survival status and should promote apoptosis. Experiments with antifreeze proteins have shown that exposure to these agents before freezing enhances the destructive effects of cold injury. During freezing, intracellular ice crystals develop at high subzero temperatures and cause damage, most likely mechanical in nature. Experiments *in vivo*, injecting antifreeze proteins into the tumor before freezing, show enhanced destruction from the use of the adjunctive agents. Considerable work needs to be done on the use of adjunctive agents, defining their role and efficacy, because all such agents are capable of causing cell destruction when used alone.

Another type of putative adjunctive therapy is defined in the concept of cryoimmunology, that is, the possibility that freezing tissue *in situ* will elicit a beneficial immunological response, as first suggested by Shulman et al. in the mid-1960s. In the following years, many studies have been directed at the nature of the immunological response, which certainly is complex. As cytokines are released in massive amounts, the reaction may lead to failure of organ systems, called "*cryoshock*" by some investigators.

On the other hand, benefit in tumors in experimental animals have been shown by some investigators and denied by others and enhanced metastases due to immunologic response have been described. So the question of benefits due to an immunological response remains unsettled. Nevertheless, immunological-enhancing drugs have been considered helpful in advanced cancers, so it is evident that the last word on the subject has not been written.

Throughout history, advancements in the field of cryosurgery have closely followed technological innovations in the handling and delivery of effective cryogens. These strides have enabled the cryotreatment of almost every known tissue type in the human body. Unique qualities (e.g., minimally invasive, allows nerve regeneration, allows treatment of irregular tissues difficult to resect, etc.) have made cryosurgery an attractive alternative for surgeons. Considerable progress in some aspects of cryotherapy was made in the 1990s. Certainly little change has occurred in the uses of cold to ameliorate pain and swelling in traumatic injuries, but cryotherapy using freezing techniques has provided benefits in diverse diseases.

Cryosurgical apparatus and imaging techniques have evolved to the state in which selective preservation and destruction techniques are practical. Research will further development in these new directions. Though researchers have established the impact of the sheer physical damage of ice-rupture, questions still remain regarding the molecular mechanisms of cell death due to the gamut of hypothermic changes that occur during cold exposure.

The elucidation of cell type-specific sensitivity and death mechanisms in association with cold exposure will allow surgeons to optimize cryosurgical procedures. In the treatment of tumors, adjunctive therapy by diverse agents promises to increase the efficacy of cryoablative techniques. In these many uses of cryotherapy, a better understanding of freezing injury, including the recognition of apoptosis, will facilitate treatment along molecular-based cellular mechanisms. The research needed to improve cryotherapeutic techniques and applications is in progress.

10

Application of Preservation

An accurate assessment of the efficacy of a biopreservation solution and protocol ultimately resides on the quality and reliability of the viability and functional assays used to assess the preservation process. The field of biopreservation viability assessment is continuing to evolve. Early in the history of the discipline red blood cell preservation was monitored using simple spectrophotometers that measured hemoglobin leakage from these cells; whereas sperm preservation was often scored based on motility. Now, however, there are a variety of assessment assays that can track a multitude of changes in cell physiology and function that either reflect cell death or herald the initiation of a key death cascade precipitated by poor preservation procedures. Yet with dozens of assays now available, a quandary is now presented on which assay(s) to choose to assess preservation efficacy. This article reviews the current slate of viability assays that have been organized into four assay tiers. The current state of the art indicates that multiple endpoint fluorescence assays hold the strongest position as a battery of viability assessment tools.

Nearly all of these viability assays can be broadly applied for most cell types, whereas functional assays typically need to be tailor-matched to the particular cell type being evaluated. A review of the literature indicates that (1) no single assay should be used to assess viability or function of cells undergoing a preservation process, (2) the unique behavior of a particular viability probe must be considered in the context of each cell type (e.g., primary hepatocytes will detoxify many fluorescent probes leading to erroneous conclusions), (3) new noninvasive assays should be developed to adequately determine the preservation efficacy of thick tissues or whole organs, (4) the future of biopreservation viability assays will be driven by a mechanism approach that may ultimately lead to the use of proteomic and genomic profiling of preserved tissues, and (5) a recommendation is posited arguing that the best method of comprehensively assessing preservation efficacy is to choose at least one assay from Tiers 1, 2, and 3 as defined herein. Thus, the portfolio of assays that is emerging is more reliable and constitutes an array of assays that should be used as a multiple endpoint approach. Furthermore, these assays are now yielding more information about the stress and cell death pathways in cells that are being activated as a consequence of biopreservation. This information, in turn, will yield improved preservation solutions and protocols in the future to support the cell therapy, gamete preservation, and tissue and organ preservation arenas.

Evolution of the Multiple Endpoint and Mechanism-based Assay Concepts

A bewildering array of viability assays that can be used to assess the efficacy of a new preservation protocol or solution formulation is now available to the preservation biologist. Some of the earliest assays of historical interest were limited to a single cell type. For instance, Gulevsky et al. used one of the earliest assays reported in the literature when they examined the effects of hypertonic solutions and slow freezing to a variety of temperatures on the permeability of plasma membranes in erythrocytes. In this case they examined hemoglobin leakage and the release of ^{14}sucrose. While these assays appeared to work reasonably well, the former is cell specific and the latter shares the challenge of necessitating appropriate radioactive disposal processes. Furthermore and fundamental to this review article, both assays measured the same parameter — the loss of membrane integrity. Thus, by using them together these assays were not designed to reveal multiple aspects of cell death that can occur as a consequence of preservation. Yet at this point in time there were few other probes or protocols that could reveal the underlying causes of preservation-induced cell death other than the breakdown of cell membranes.

The next group of assays that appeared in the literature was essentially improvements and extensions of the same type of release assays as those mentioned above. The ^{51}chromium (Cr) release assay functioned in a manner similar to the hemoglobin and glucose release assays. Unlike the glucose release assay, the ^{51}Cr release assay could be used with a multitude of cell types. This assay was used actively in the 1970s–1990s to assess preservation efficacy. For instance, Holden et al. used the ^{51}Cr release assay to study the cryopreservation of lymphocytes. Callery et al. also used the ^{51}Cr release assay to study the cryopreservation of lymphocytes and correlated the release of chromium to rosette-forming cells. The most recent reference to this release assay was that of Schmidt-Wolf et al., who used it to assess the cryopreservation of murine lymphokine-activated killer cells. By this time, however, the use of nonradioactive probes that could substitute for radioactive isotopes was in vogue and preservation biologists began to champion other viability assays such as the LDH release assay.

The LDH (lactate dehydrogenase) assay appeared in the literature at about the same time as the ^{51}Cr release assays were exiting the laboratory. Once again, the LDH release assay was limited to measuring cell membrane leakage. Yet unlike the ^{51}Cr leakage assay, the LDH assay is still used today, given that it (a) is nonradioactive, (b) varies little between different cell types, and most importantly, (c) can be used both on single-cell monolayers as well as to assess whole organ preservation. For instance, Ostrowska et al. used the LDH release assay to examine the functional and morphological integrity of freshly isolated and cryopreserved human hepatocytes. Not surprisingly, they found a tight inverse relationship between LDH leakage and trypan blue exclusion given that both assays measure cell membrane breakdown.

As stated previously, however, the LDH leakage assay is still used today, given that it is an assay that works well examining the post-thaw function of hypothermically stored organs such as liver. Tan et al., for instance, used this assay to examine the protective effects of exogenous adenosine triphosphate on hypothermically preserved rat liver. Quintana et al. examined the effect of S-nitrosoglutathione added to the University of Wisconsin (UW) solution and were able to correlate the release of LDH in hypodermically stored livers to glycogen content, the release of bile, and the morphological appearance of sinusoidal endothelial cells. Finally, a group in the Netherlands compared the efficacy of UW solution to HypoThermosol using the LDH assay and found both solutions to be comparable and once again, the release of LDH appeared to correlate well with liver enzyme function. Thus, the LDH leakage assay is still commonly used today given its flexible applications ranging from single cells to whole organs.

The late 1980s ushered in a revolution in how *in vitro* toxicity assays can be accomplished with cells seeded in multiwell plates.

Two simultaneous events occurred at this time. First, the then-young company Molecular Probes of Eugene, Oregon, introduced the power of fluorescent probes to monitor a variety of cell functions by offering hundreds of indicator dyes primarily designed to be used with either a flow cytometer and/or the relatively new, confocal microscope. Yet, the implementation of fluorescent dyes to measure cell toxicity was at the outset limited to those facilities that had these expensive instruments in house. In both cases, neither the FACS nor the confocal was designed for high throughput screening. In the late 1980s, however, Millipore Corporation introduced the first bottom-reading spectrofluorometer (CytoFluor) expressly designed to read fluorescence *in situ* in monolayers of cells grown in multiwell plates.

With the ever-increasing number of physiology-specific fluorescent indicator dyes now marketed by a variety of vendors and the advent of this new type of bottom-reading spectrofluorometer that made high throughput screening possible, it was now much easier to accomplish a variety of fluorescence-based assays often referred to as a battery or multiple-endpoint assay. The term "endpoint" in this phrase marked the beginning of being able to look at multiple cytotoxicity indicators that reflected different endpoints of cell death. Thus, the investigator could look at cell membrane disruption, loss of mitochondrial proton motive force, influx of intracellular calcium and other parameters that reflected different endpoints other than merely the ultimate loss of cell membrane integrity that all cells experience once they have gone through the cell death process.

Furthermore, given the breadth of fluorescent probes emerging it was possible to do double and triple labels of the same cells so that multiple parameters could be measured at the same time and in the same cells. Thus it was clear by the early 1990s that fluorescent probes had the versatility, chemical specificity, and spectral qualities to be considered the next generation of viability indicators that could be used to assess preservation efficacy. This multiple-endpoint concept is echoed in the current online Molecular Probes Handbook that states the following — Overview of Probes for Cell Viability, Cell Proliferation, and Live-Cell Function: "The diversity of live cells and their environments makes it impossible to devise a single viability or enumeration assay applicable to all cell types. Because viability is not easily defined in terms of a single physiological or morphological parameter, it is often desirable to combine several different measures, such as enzymatic activity, membrane permeability, and oxidation-reduction (redox) potential. Each assay method has inherent advantages and limitations and may introduce specific biases into the experiment; thus, different applications often call for different approaches."

The third important stage in post-thaw viability assessment occurred more recently when Baust et al. recognized that by using a mix of both the fluorescent multiple endpoint assays and spectrofluorometer analysis of enzymatic assays, it was possible to determine not only if cells were succumbing to a preservation process, but in addition it was possible to determine *how* this occurred and to plot a time frame in which cell death due to extended or poor cryopreservation is executed. A critical contributor to this next level in preservation assessment was the introduction of Alamar-Blue, a metabolic indicator dye reviewed later. This probe allowed the investigator to assess the metabolic activity of the same cell culture repeatedly over a nearly unlimited time period. This near-real-time assay was critical for the Baust group to identify delayed-onset cell death. These data, reviewed later, have led us to the understanding of how preservation-induced apoptosis and its prevention are important considerations in preservation protocols.

Thus, viability assays used in the preservation sciences can be seen as three historical chapters marked by (1) the early membrane lysis studies, (2) the multiple endpoint assays that measure different cell parameters, and (3) the more mechanistic assays that have now revealed the importance of both apoptosis and necrosis in preservation-induced cell death. It is the intent of this article to review most of the assays mentioned in the current literature and organize them in a multiple tier level so that a recommendation can be made for a multiple endpoint assay that can comprehensively assess preservation efficacy and molecular mechanisms. An example of how these powerful assays can be used in concert together as a step toward improved preservation protocols and solutions is presented elsewhere. This Multiple Solution Hypothesis, a new concept in the preservation sciences, would not be possible without the appropriate mix of assays from Tiers 1, 2, and 3 listed in the next section. It is also the intent of this article to look at new and emerging technologies such as cDNA microarrays and SELDI-TOF/ProteinChips and to make projections on how these and other techniques will be important in the future as assay systems to understand and improve preservation formulations and protocols.

Multiple Assay Tier Concept of Viability Assays

A survey of nearly all viability assays available to the preservation scientist reveals that they can be grouped into at least four categories.

The following assay tier is not specific to preservation biology, but is presented below as a tier concept that can guide those who work with preserved cells.

Tier 1 Assays: Cytolysis Live/Dead Assays

This tier of assays is historically the first to be used to assess viability and still remains a key group of techniques that should be included in all preservation viability assays. The chromium and glucose release assays reviewed earlier are of historical importance and thus are not reviewed here, but do represent cytolysis or membrane leakage assays and thus are included in this assay tier. The cytolysis assays have both a very positive and a negative attribute to them. On the positive side, there are a variety of assays that can reveal cell membrane leakage that occurs as a final stage in most forms of cell death. Thus, the investigator has the flexibility to use these assays coupled to a microscope, spectrophotometer, and/or a spectrofluorometer. Yet given that cytolysis is the last stage of preservation-induced cell death, these assays do not reveal early-stage mechanisms underlying preservation-induced cell death and thus have limited use in the future as a diagnostic means to develop improved preservation formulations and protocols.

The LDH assay, unlike its predecessors, continues today to be useful for measuring preservation-induced cytotoxicity. The concept behind this cytolysis assay is simply that if the cell membrane is compromised, then LDH will leak into the extracellular milieu where it can be measured. LDH has at least three advantages over alternative enzymes that could have been selected as candidate enzymes to be measured. First, LDH is a common enzyme in all cells and varies little with the metabolic state of the cell. Second, unlike many other cytoplasmic enzymes, LDH exists in a relatively high concentration and is stable. Third, LDH can be measured using a coupled enzymatic reaction. In this series LDH oxidizes lactate to pyruvate which, in turn, can be converted to form formazan via the trazolium salt, INT. While there are a number of variations on how this can be accomplished as revealed in several protocols, the end result is that the LDH is measured indirectly via a spectrophotometer that measures formazan at an absorbance of 490–500 nm. There are several companies that offer protocols and kits that give detailed information of how to assess LDH leakage. Surveying these protocols as well as original research articles has shown that the LDH assay is reliable, but there are at least two drawbacks to this assay. First, several different controls must be executed to compute the results. For instance, BioVision lists a background control, low control and high control; whereas the CytoScan LDH-Cytotoxicity Assay Kit suggests employing seven different controls. The second drawback is that there are relatively few viability assays that use spectrophotometric analysis, whereas the more sensitive and utilitarian spectrofluorometry and bioluminescence are now more preferred. The LDH leakage assay does, however, have an important virtue not shared by most other assays. As mentioned previously, it can be used to assess the preservation efficacy of both single cells as well as whole organs. For this reason, the LDH leakage assay should be considered a key assay in Assay Tier 1.

The trypan blue assay is also one of the most commonly used cytolysis assays. A number of investigators has used the trypan blue exclusion assay in studying preservation efficacy. For instance, Abrahamse *et al.* used the trypan blue assay to measure necrosis as one of many assay tools to compare the hypothermic preservation efficacy of UW, HTK, and Celsior solutions. Muller *et al.* used the trypan blue procedure to assess serum-free cryopreservation of porcine hepatocytes. Finally, Isayeva *et al.* did an interesting study on the chilling sensitivity of zebrafish oocytes and compared the sensitivities of the MTT assay to the trypan blue assay and found the latter more sensitive. It should be noted, however, that the MTT assay measures mitochondrial activity that may or may not be related to cell membrane leakage measured by the trypan blue assay.

The principle behind the trypan blue assay is simple. This dye is negatively charged and will bind to positively charged proteins in the cytosol only if the plasma membrane is ruptured. The advantage of the trypan blue assay over the LDH leakage assay is that the former is much easier to execute, yet demands the use of a hemocytometer to count trypan blue-stained cells. Furthermore, fewer positive controls are necessary to run. For instance, LDH exists in serum and thus can influence results when one is testing cells preserved in serum-based preservation media. The presence of serum has little effect, if any, on the trypan blue assay. It does, however, share the same handicap as the LDH assay given that neither can be analyzed using fluorescence and/or bioluminescence. Thus, in both bases, the LDH and trypan blue assays are not amenable to high throughput analysis. For additional information on the trypan blue exclusion assay, more details can be found in Freshney's *Culture of Animal Cells: A Manual of Basic Technique*.

Clearly, the future of Tier 1 assays is in bioluminescence and fluorescence, given that both are now being more commonly used in cell and molecular biology protocols. Of these two, fluorescence dominates. Yet one particular assay that uses bioluminescence is worth mentioning, although it has not yet been used for assessing preservation efficacy. The ToxiLight BioAssay kit works via bioluminescence. It measures the release of adenylate kinase from cells whose cell membranes have been damaged. Like LDH, adenylate kinase is present in all eukaryotic cells. This enzyme converts ADP to ATP and it is the latter that is measured using the bioluminescent firefly luciferase reaction. Thus, as the membranes break down more adenylate kinase is released from the cells. The virtue of this assay is that it is one of the few that can be used over time with the same cell culture set and is amenable to high throughput analysis. For instance, a robotic system can be employed to collect very small samples of media on a regular basis to monitor the release of the enzyme. Given that it uses the luciferase enzyme, it has a range of over 3 orders of magnitude and can detect as few as 10 cells/microwell. As such, it is a non-destructive or noninvasive assay that can reveal small changes in cell membrane permeability. According to the manufacturer, it requires only one step and results can be generated in 10 minutes. Thus, this assay appears to have great potential for assaying preservation-induced, time-dependent changes in cell membrane integrity.

The more commonly used set of Tier 1 assays are those that employ fluorescent indicator dyes. The explosion in the use of these fluorescent dyes as well as the plethora of multiwell-reading spectrofluorometers is a testimony to the utility of these indicator dyes. A large number of preservation investigations have been accomplished in the past few years that have employed Tier 1 fluorescent probes that include Calcein-AM, CFDA-AM, BCECF-AM, SYTO, ethidium homodimer, and propidium iodide. This group of probes can be subdivided into two different subsets, one of which is trapped by the cell and leaks out only if a membrane rupture occurs. An example illustrated later in this article is Calcein-AM. The other subtype, exemplified by propidium iodide, is membrane insoluble and only stains the cell if it gains access through a compromised plasma membrane.

Most of the probes in the first category contain an acetoxymethyl ester ("AM") that confers membrane permeability to these dyes. As electrically neutral molecules, the esterase substrates have been very useful for loading cells with a variety of indicators that can measure intracellular ph (BCECF-AM) to calcium (FLUO3-AM). In the hands of this investigator as well as corroborated elsewhere, Calcein AM is the dye of choice within this Tier 1 group due to its superior retention, resistance to change in emission intensity due to changes in pH, and the fact that it is read at fluorescein wavelengths and thus can be quantitatively and qualitatively analyzed using standard spectrofluorometry and fluorescence microscopy. BCECF-AM, on the other hand, is also available to be used as a Tier 1 assay, but its emission intensity is only half-maximal at pH 7.0. Fluorescein diacetate was one of the first probes to be used to monitor cell membrane lysis because fluorescein is formed by intracellular hydrolysis of fluorescein diacetate, yet fluorescein readily leaks from cells. This high leakage rate led to the development of carboxyfluorescein diacetate (CFDA) which upon hydrolysis forms carboxyfluorescein, but Calcein-AM is regarded as the premier fluorescent dye that measures plasma membrane integrity.

A number of preservation biologists have used Calcein-AM, propidium iodide, and CFDA to investigate improved preservation protocols and solutions. Grundler et al. used a SYBR-14/propidium iodide stain combination to identify viable, dead, and "intermediate" bull sperm subpopulations subsequent to cryopreservation; whereas Liu et al. used propidium iodide in conjunction with Annexin V to look at the anti-apoptotic effect of ascorbic acid after cold reperfusion injury in rat liver. Sion et al. also used the Annexin V/propidium iodide combination to correlate spermatozoa viability to function. Calcein-AM has also been used by a variety of laboratories, including our group, to assess cell membrane leakage as a consequence of cryopreservation.

For instance, the comparative efficacies of hypothermic preservation solutions have been assessed by our group using Calcein-AM and other fluorescent indicator probes. Neonatal human hepatocytes were stored in a variety of hypothermic storage solutions for two days and allowed to recover at normothermic temperatures for one day. At this time cells were incubated with Calcein-AM, washed, and analyzed in both the fluorescence microscope or using the CytoFluor multiwell plate reader. Note that Calcein-AM is able to distinguish differences in the abilities of the hypothermic preservation solutions to protect the cell membrane integrity of these hepatocytes. Baust et al. have also shown that Calcein-AM can be useful to demonstrate the delayed- onset cell death that occurs to cells during the 48 hours subsequent to thawing,

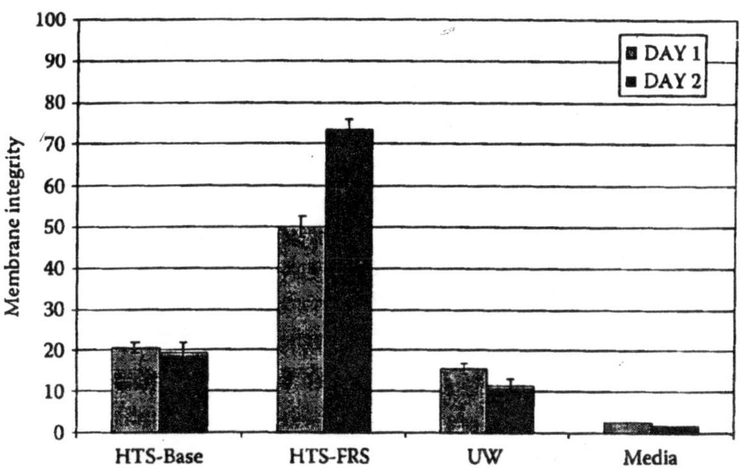

Fig. 10.1. Hypothermic storage of neonatal hepatocytes following 2 days at 4°C.

and can be combined with Annexin V and propidium iodide to yield estimates of cells succumbing to necrosis and apoptosis. Thus, Calcein-AM and the other associated plasma membrane integrity dyes have been extensively used as a first tier to assess cell viability of preserved cells through analyzing cell membrane integrity.

Tier 2 Assays: Enzymatic and Molecular Mechanism Assays

The Tier 2 assays are designed to yield more detailed information of the basis underlying cell death that can occur as a consequence of cell preservation. In most, if not all, cases, these assays reflect changes in cell physiology that can occur that precede the later event of the loss in cell membrane integrity. Once again, given the emphasis on fluorescent indicator dyes in this chapter, the major focus will be on this type of probe. Most of these indicator dyes focus on either events that are components of, or linked to, apoptosis — programmed cell death. The other group of probes indicate changes in mitochondrial activity that may or may not be linked to apoptosis and/or necrosis that occurs as a consequence of preservation-induced cell death. Given the breadth of probes available from several vendors, it is beyond the bounds of this discussion to list all of these assays, but the probes used most often in the preservation sciences will be discussed.

Mitochondrial probes

A number of probes have been very useful in assessing mitochondrial activity. One of the most commonly used probes is the resazurin derivative, alamarBlue. Other mitochondrial probes include the tetrazolium salts that have been used for analyzing redox potential of cells for viability, cytotoxicity and proliferation. These compounds form a nonfluorescent formazan product and thus cannot be incorporated easily in a multiple endpoint fluorescence assay system. Yet the reduction of MTT is one of the most popular mitochondrial activity assays used today for a variety of applications and can be quantified using standard absorbance readers.

Castagnoli et al., for instance, used MTT successfully to correlate changes in structure and changes in viability in cryopreserved skin. Son et al. has also used the MTT assay for optimizing cryoprotectants for the cryopreservation of rat hepatocytes. On another front, a large group of dyes have emerged that are regarded as potentiometric probes. These dyes, like their parent compound Rhodamine 123, are mitochondrial selective and can be used to analyze the proton motive force of mitochondria. In most cases the amount of Rhodamine 123 retained by the mitochondria reflects the magnitude of the proton motive force. While once again the diversity of these potentiometric probes is too extensive for a full review, many of these dyes, such as JC-1, have been found extremely useful for examining changes to mitochondrial potential as a consequence of preservation.

This chapter will focus briefly on alamarBlue and JC-1, given that both have been used to assess preservation efficacy. AlamarBlue is a noninvasive fluorescent dye that can be analyzed either spectrophotometrically or spectrofluorometrically. While our group has used this dye extensively for both cryopreservation and hypothermic preservation studies, the reader is best directed to an article by Voytik-Harbin et al. that compares alamarBlue to the MTT and radioactive thymidine assays alamarBlue has several virtues. First, as demonstrated by the aforementioned group and verified by our research team, the alamarBlue assay is a superior assay for counting cells *in situ*. The greater the fluorescence signal, the larger the number of cells being measured. Second, while most investigators analyze alamarBlue using a spectrofluorometer, the conversion of alamarBlue to its red alternative can be tracked visually.

Thus, in much the same manner as developing black and white photo prints in a developer solution, the investigator can subjectively determine the optimum incubation time that will demonstrate differences between samples. But alamarBlue's major virtue lies in the fact that it is nondestructive and can be used repeatedly with the same cultures over an extended period of time. This has been especially important to our research team given that it has allowed us to track the time-dependent decrease in cell viability that can occur subsequent to preservation. Human keratinocytes were seeded in cell culture plates, subjected to hypothermic exposure while immersed in a variety of hypothermic preservation agents, and alamarBlue fluorescence (expressed as "*percent viability*") was analyzed over a three-day period. Note that the alamarBlue fluorescence of cells preserved in UW solution was highest at Day 0 compared to the HTS series. Yet over the course of 3 days the apparent metabolic activity increased in the HTS series, whereas it declined in cells stored in either UW solution or KGM (keratinocyte growth media). Thus, if this experiment had been limited to Day 0, then the assumption might be that the UW preservation solution is superior to HTS; yet the ability of alamarBlue to analyze the same cultures repeatedly over a long period of time showed that the apparent metabolic activity of the cells stored in UW appeared to decrease over time. Given that changes in alamarBlue can reflect metabolic activity and/or an increase in cell number, additional experiments not shown here using SYTO dye (cell membrane permeable nuclear stain) and Calcein-AM showed that these changes reflected an increase (HTS) and a decrease (UW, KGM) in cell number, respectively. Thus, the alamarBlue assay was instrumental for our group to demonstrate that preservation-induced cell death can take several days to be manifested.

As mentioned previously, a number of mitochondrial potentiometric dyes are available for analyzing viability. In contrast to alamarBlue that is quantified while present in a simple, extracellular, assay medium, the aforementioned group of dyes are *in situ* dyes that localize to the mitochondria. Each has its own virtues and limitations. Rhodamine 123, for instance, is a cationic dye that localizes to the mitochondria but often binds nonspecifically and creates high background problems. Also, this and other cationic mitochondria dyes can be retained by some cells for only an hour or two, whereas other cell types such as transformed cells can retain this dye for as long as a week. Thus, the behavior of this dye does not lend itself to routine application in the preservation sciences. Recently, however, our group has tested the JC-1 mitochondrial dye to determine changes in mitochondrial activity in response to different preservation protocols and solutions.

JC-1 (5,5',6,6'-tetrachloro-1,1',3,3' -tetraethyl-benzimidazolyl-carbocyanine iodide) can exist either as a green monomer or as a red fluorescent "J-aggregate." The green emission profile represents mitochondria with relatively low proton motive force; whereas the red emission represents mitochondria with robust and healthy proton motive force. The important nature of this dye to preservation biologists is that it can directly analyze the mitochondrial activity of cells that have emerged from a preservation sequence and the time-dependent recovery of mitochondria can be tracked in replicate samples. Martinez-Pastor et al. have used JC-1 to assess the cryopreservation of ram semen and it constituted one of a battery of assays that measured different viability and functional endpoints. Pena et al. also used JC-1 and correlated emission profiles to sperm motility. Our research team hypothermically preserved normal human dermal fibroblasts in either ViaSpan or HTS-FRS, returned the cells to normothermic temperatures, and then analyzed them using JC-1. The data illustrate that JC-1 formed monomers in the mitochondria in cells preserved in ViaSpan, whereas the dye formed the J-aggregates in the mitochondria in cells preserved in HTS-FRS. Thus, the mitochondria in cells preserved in the latter preservation medium appear to have a higher metabolic rate than those in cells preserved in ViaSpan. This dye is also being used in other venues to analyze the mitochondrial permeability transition pore — a complex system in the mitochondrial membrane that may also be important to consider in the quest to improve preservation solutions and protocols.

Apoptosis probes

One area in the preservation sciences that has been relatively new to appear in the science literature is the apparent ability of both hypothermic preservation and cryopreservation to induce apoptosis or programmed cell death. Cook et al. noted that the extended storage of engineered tissues in any type of preservation medium can result in a time-dependent, incremental decrease in cell or tissue viability that can last up to one week's time. In this experimental series Cook et al. incubated the engineered human epidermis, EpiDerm, for a variety of times in hypothermic preservation solutions. They determined that using alamarBlue, the tissue viability could be monitored for a week or longer subsequent to hypothermic preservation. The data show that the viability of EpiDerm stored for 11 days, however, declined each day subsequent to return to normothermic temperatures. The histology of the tissue reflected the apparent drop in viability as seen by the daily alamarBlue readings. Yet it was not clear until the work done by Baust et al. several years later (1999) that this preservation- induced

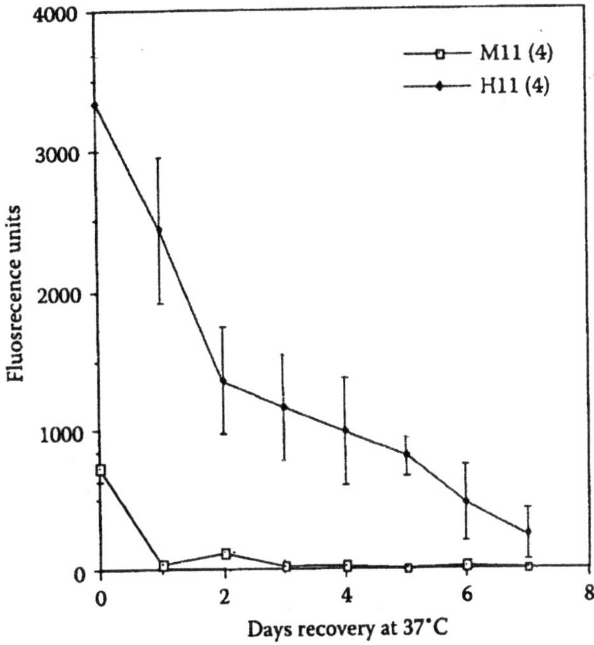

Fig. 10.2. Hypothermic preservation of EpiDerm in HTS-BASE.

decline in viability referred to later as "delayed onset cell death" was due to apoptosis. These observations, coupled with the increasing importance of apoptosis as the keystones to developing new treatments for cancer, launched a new era of assays available to the preservation biologist.

A variety of tools that range from DNA gel electrophoresis to fluorogenic assays can now be used to study preservation-induced apoptosis. DNA gel electrophoresis has been used by our group to study both apoptosis and necrosis in cryopreserved cells. For instance, in one such case, MDCK cells were hypothermically preserved for various periods of time and then returned to normothermic temperatures. The pattern of the gel suggests that apoptosis is a major contributor to preservation-induced cell death, whereas necrosis is more important at later times. Thus, it is clear that DNA gel electrophoresis can be used as a later-stage assay to determine how cell death occurs to cells that have been subjected to extended or faulty preservation.

A variety of assays are now available that can examine events that are upstream of DNA cleavage. While there may be a number of ways that poorly preserved cells can launch apoptosis (e.g., activation

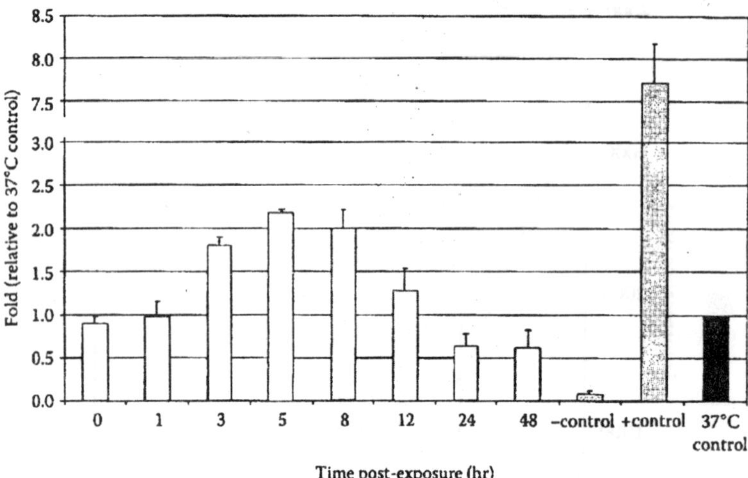

Fig. 10.3. Increase in caspase-3 activity in CAEC subjected to cryopreservation.

of cell death receptors, opening of the mitochondrial permeability transition pore, etc.), most of these involve the activation of caspases — the executioner enzymes of apoptosis. Baust et al. were among of the first to show that cryopreservation caused an increase in caspase activity — a process that can be analyzed by using a fluorogenic substrate kit or through traditional RT-PCR. Since this time a variety of new methods have appeared that allow one to analyze the activation of caspases. R&D and other groups offer caspase fluorometric assays that can be used to examine preservation-induced caspase activation. Cells are typically subjected to a preservation episode, the cells are then lysed, and the protease activity determined via the addition of a caspase-specific peptide that is conjugated to the fluorescent reporter molecule, 7-amino-4-trifluoromethyl coumarin. The cleavage by the caspase of this peptide releases the fluorochrome that can be subsequently detected in a spectrofluorometer. The activity of the caspase is directly related to the fluorescence intensity of the sample. Note that there was a time delay in the activation of caspase activity subsequent to return to normothermic temperatures.

Other fluorogenic assays are now available that can detect active caspases *in situ*. For instance, Molecular Probes offers a host of cell-permeant caspase affinity labels that can be used in this manner. Their Vybrant FAM caspase assay kits are designed for flow cytometry whereas their new Image-iT LIVE caspase detection kits focus on microscopy. Both use a fluorescent inhibitor of caspases (FLICA) that can bind covalently and irreversible to active caspases and unbound FLICA can be easily washed away. BioVision offers CaspGLOW fluorescein caspase staining kits that can also detect the *in situ* activation of caspases in living cells. Like the Molecular Probes kit, CaspGLOW consists of the caspase family inhibitor, VAD-FMK, conjugated to FITC that irreversible binds to active caspases in living cells. As of yet, no published reports have appeared that document the use of either of these *in situ* dyes for studying preservation-induced apoptosis.

Finally, one of the well-documented apoptosis probes that is an excellent example of the power of multiple fluorescent probes is the commonly employed Annexin V/PI dual stain. Propidium iodide was discussed previously and is a Tier 1 assay. Annexin V can be regarded a Tier 2 apoptosis assay because it measures the flip-flop of phosphatidylserine from the inner plasma membrane leaflet to the outer phospholipid layer where it can be detected by externally added, membrane-impermeable, Annexin. This inversion of phosphatidylserine is one of the earliest events in apoptosis and has been used by our group to study cell death events occurring as a consequence of both hypothermic storage, as well as cryopreservation. It is noteworthy that this double-label technique produces four distinct populations. One group of cells does not stain to any extent with either PI and/or Annexin V. A second population is one that stains exclusively with Annexin and thus represents cells that have probably launched the apoptosis cascade. A third population is one that stains exclusively with PI and may represent the necrosis group. It is the fourth group, that is the most problematic. These are cells that stain with both Annexin V and PI and, as such, may be undergoing secondary necrosis. Secondary necrosis defines the time in the cell death cycle where the cell has undergone apoptosis but the cell membrane is permeable enough so that the nucleus stains with PI. This is a critical matter to consider given that it is important to do a series of timed studies subsequent to thawing of preserved cells to monitor both the PI and Annexin V-only staining. Double-labeled cells may indicate that the early events of apoptosis may have already occurred.

Miscellaneous

Tier 2 assays are defined herein primarily as those that either measure mitochondrial activity or assess early stage and late stage apoptosis/necrosis events. There are, however, many other contemporary assays that can be used to monitor changes in cell physiology that lead to apoptosis, necrosis, and mitochondrial dysfunction events. For instance, Molecular Probes offers MitoSOX Red reagent and Image-iT LIVE Green Reactive Oxygen Species Kit that can monitor oxidative stress in live cells. While useful for mechanistic studies, these assays should not be considered viability probes. Yet these assays, in particular, are important in assessing the oxidative stress that occurs to cells, tissues, and organs subsequent to either a cryopreservation procedure or a hypothermic storage episode.

Tier 3 Assays: Functional Assays

One of the key components of any preservation protocol must be a functional assay that matches the type of cells or tissue being analyzed. In some cases this determination is quite easy. For instance, sperm motility is the method of choice for sperm cryopreservation. Martinez-Pastor et al. showed that there was a correlation between mitochondrial activity (JC-1) and ram sperm motility — a correlation that is not surprising given the link between the two. Pena et al. showed that the addition of the antioxidant trolox increased both JC-1 staining and motility of boar sperm. Thus, these and other studies suggest that there may be a relationship between the viability and functional assay. Finally, Nallella et al. used a motility assay to evaluate two cryopreservation methods and three cryoprotectants on the preservation of human spermatozoa. Thus, the sperm motility is a well-accepted functional assay for this system.

The choice of a functional assay for the hypothermic preservation of single cardiomyocytes is also straightforward. Our group has compared a number of different hypothermic storage solutions that includes cell culture media, ViaSpan, and a number of Hypothermosol variants for their abilities to protect cardiomyocytes at 4°C. In this case cultures were scored for the percent cells beating and the same cultures were then assessed for viability using Calcein-AM. Note that there is a tight relationship, once again, between the viability assay and the functional test.

Finally, the preservation of livers and hepatocytes is an active area of research and in most cases functional assays are accomplished using a battery of assays. An experiment was designed to determine if the addition of fetal bovine serum to HTS-FRS improved the performance of this solution compared to HTS-BASE (a different HTS variant) or HTS-FRS without the serum supplement. In this case five different drug-metabolizing enzyme assays were performed (CYP3A4 =testosterone 6β-hydroxylation; ECOD = 7-ethoxycoumarin Odeethylation; FMO = p-tolyl methyl sulfide oxidation; UGT = 7-hydroxycoumarin glucuronidation; ST = 7-hydroxycourmarin sulfation). It should be noted that there was a similar functional efficacy order in most of the assays. For instance, all five assays suggested that the hypothermic performance of HTS-FRS was superior to that of HTS-BASE supplemented with FBS. It is curious that the data also may suggest that storage in HTS-FRS and HTS-FRS-FBS may improve or restore the function of the liver cells compared to freshly isolated cells ("Isolation (0 hr)"). It should also be noted that the viability of the liver cells stored in HTS-FRS was superior to that of cells stored in HTS-FBS (data not shown). In this case viability assays were restricted to Calcein-AM given that primary hepatocytes can detoxify the alamarBlue.

Thus, in the case of the preservation of sperm, cardiomyocytes and liver cells there appeared to be correlations between viability and function. Yet this is not always the case in other systems. Neonatal human keratinocytes used to construct the epidermis that is a case in point. Our research team has noted that a single strain of these cells can be cryopreserved multiple times and both cell proliferation and viability do not appear to be grossly compromised as a result of these multiple preservation procedures. Yet when these cells are placed on the appropriate microporous membrane system designed so that the cells can stratify and differentiate into the engineered construct, cells cryopreserved multiple times are unable to differentiate fully and form a stratified epidermis. This observation implies that multiple cell and tissue-specific functional assays should be considered carefully for human stem cells cryobanked for subsequent use in cell therapy applications.

Tier 4 Assays: Genomic and Proteomic Assays

Tier 4 assays consist of a group of protocols that have not yet been tested extensively by preservation biologists but may prove to be useful in the future design of improved preservation solutions and protocols. Both cDNA microarrays and protein chips are available through a number of vendors that can analyze preservation-induced stress pathways as well as examine the protein profile of cells that have been subjected to hypothermic storage or cryopreservation. Our group has tested Ciphergen's SELDI ProteinChip system for its ability to distinguish changes in cells subjected to different cryopreservation protocols. Clear differences can be noted in the profiles, yet it is unclear at this time how these data can be used to predict viability, function, or both. Yet these assays will be crucial to the advancement in preservation biology given that they have the capability of determining which of the stress pathways are activated in cells. Once known, designing improved protocols or solutions that can suppress these pathways may lead to improved procedures for cell preservation.

A number of assays are available to the preservation biologist. Fluorescence probes have now proven to be the best choice for accomplishing viability and functional assays given their abundance, versatility, and high sensitivity. New instruments, such as Molecular Devices, FLIPR system designed for high throughput analysis, may allow preservation biologists to look at multiple fluorescence probes in a near-real time format so that changes in cell physiology (e.g., mitochondrial potential, intracellular calcium, etc.) can be followed over the course of several days subsequent to return to normothermic temperatures. Furthermore, it is important that multiple assays be chosen to assess viability and function as long as they are not measuring the same physiological parameter. This problem can be avoided if assays are chosen from Tiers 1, 2, and 3 as described in this article. Finally, there needs to be a focus on the Tier 4 assays that are more diagnostic in nature so that improved preservation protocols and solutions can be developed.

INDEX

U

V

W